So What Happened *to* God?

Renee...

It is a complete joy getting to truly know you. You embody so much of what is important about being human. You are curious, interested in others and eager to be involved. You are so loving and playful & eager to jump in. You even have compassion for the much maligned creatures called "fathers"! ☺

You & your work in the world are a huge blessing for sure, more than you even can imagine!

Love
Bruce

So What Happened to God

Religion, Science, and Democracy?

Poems & Essays by

Bruce Silverman

REGENT PRESS
Berkeley, California

[paperback]
ISBN 13: 978-1-58790-469-1
ISBN 10: 1-58790-469-1

[e-book]
ISBN 13: 978-1-58790-470-7
ISBN 10: 1-58790-470-5

Library of Congress Control Number: 2019930202

Manufactured in the U.S.A.
REGENT PRESS
Berkeley, California
www.regentpress.net

CONTENTS

Chapter 7. The Hebrew Bible: Brief Stops Along the Way

Chapter 8. Love and Its Mysterious Manifestations

Preface

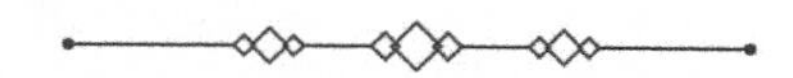

Every book project has a unique unfolding and mine is no exception. Original poetry is a relatively new adventure for me; I didn't major in literature as a college student or, as a young man, feel called to being a writer.

The essays and poems in this book came from my longtime exploration of mythology, sacred ritual, various world drumming traditions, and my return to my own Jewish religious practice after years of wandering through the waters of other mystical traditions that still nourish me.

More specifically, the book grew out of my five-year participation in the Contemplative Shabbat services at Chochmat Halev congregation in Berkeley, California. Then and currently, I am the drummer/percussionist on a team of talented musicians and ritual makers offering these services. At the helm is Estelle Frankel, a storyteller and ordained rabbinic pastor.

In February of 2014 the Bay Area was experiencing an all-too-common phenomenon, a drought. Our ritual team decided to create an evening that would attempt to break the drought through use of rain sticks, prayers, teachings and poetry. Unable to find a fitting poem in my sources, I wrote my own poem, called *Prayer for Rain,* and thus created a new role for myself.

Realize that I was never what you'd call a devout Jew; I studied more Joseph Campbell and Karen Armstrong than Hebrew Bible, so my perspectives on Torah are those of a connected outsider. I often feel like a falcon flying over the canopy of the Jewish mystical rainforest, viewing the complexities of biblical text and ritual from a distance that allows me to see these timeless stories from a universal perspective, and to wrestle with them irreverently. Many of the poems and matching essays are not connected to Judaism at all or to any religion as such. They fall more into the categories of spiritual poetry and mythological reflection.

Speaking as a musician, I hope you will feel the music in the poetic offerings and enjoy some of the literary devices herein that are less in fashion than in days past, such as rhythm, meter, rhyming, and alliteration.

I also hope that you will accept and enjoy the playfulness inherent in my critiques of what I call the *bad theologies* that are wreaking havoc on our planet. This book is my tiny way of throwing in my lot to ease a world on the tipping point of uninhabitability, and mired in the swamp of ignorance and cruelty. It is also my hope that in some fashion, what is of dire importance can still be offered with a dash of humor.

Acknowledgements

Writing a book is seldom, if ever, a solo project. When all is said and done, it's a community project. With that in mind I want to express my gratitude to my wife Audrey who supported me throughout this labyrinthine process. So often she was a valuable sounding board for words, ideas, and visions, and, she was the steadfast partner who held the larger container of our home life in countless ways so that my early morning marathons at the kitchen table could endure.

Over my lifetime, a significant number of teachers and mentors were and are noteworthy. I thank my parents, Bruce and Sylvia Silverman, for being attentive to me and encouraging me to pursue drumming, which has been a springboard into other ventures that explore rhythm and sound in esoteric ways. I am indebted to Dr. Angeles Arrien, an affirming teacher and colleague in the cross-cultural worlds of mythology. I have been nourished by the teachings of Rabbi Zalman Schachter-Shalomi; I've been mentored by Rabbi Sholom Groesberg, who carefully guided me back to a deeper exploration of Judaism by leading and helping Audrey and me in the formation of our own Havurah: called Shir Neshama, a loving community for twenty-two years now.

I thank Dr. Howard Teish for spearheading the solar-lunar perspectives that inform this book and for countless teachers along the way.

I have been encouraged by various master drummers: Zakir Hussain, Marcus Gordon, and Jose Lorenzo, and I feel gratitude to the Rev. Matthew Fox, Jim Garrison, Robert Bly, and other teachers in the worlds of theology, poetry, and social artistry. I'm grateful for the supportive allies in the men's work that have so nourished me, especially for Jay Roller, Dan Zola, and Max Lan.

In the most compelling way I am grateful to my friend and colleague Estelle Frankel, a gifted story-teller, psychotherapist and teacher of Jewish mysticism, who encouraged me to step forward with my spiritual poetry as a welcome adjunct to her Contemplative Shabbat services at Congregation Chochmat Halev.

Those services, in many ways, were the spark for the text that follows. Also, great thanks to Rabbi Brian Schacter-Brooks and his Torah of Awakening community, and to our informal Berkeley Havurah, connected to it.

I also want to acknowledge Mark Weiman, the publisher of Regent Press for his patience and guidance in this project and for my long-time friend and editor Jamie M. Forbes, for his expertise and care. I am also indebted to Vicki K. Silverman for her highly skilled and generous technical support.

Lastly, I want to honor my children, Elana Silverman and Naomi Silverman, for encouraging me to be both father and teacher, as well as therapist, drummer, parade leader, author, and a rambunctious personage in the world. Their continued support has been a blessing for me, as they are now becoming a great blessing for so many others.

Introduction

I've always been captivated by poems and essays about wells. Pablo Neruda, David Whyte, and others remind us that below the daily reality that we take for granted, there dwell other matters of import. In that sense I've been drawn to music, theology, psychology, mythology, and the psychic arts. The metaphor of the dark well touches upon these concepts succinctly. Religion's job is to remind us of these numinous waters that live beneath us. When religion is misunderstood or becomes irrelevant to a world in flux, we then encounter what some simply call: *bad theology.*

Consequently I believe that *bad theology* lies at the core of how we think, how we pray, how we treat our fellow humans, and how we care for all beings, or not, on our planet. These days we seem to have an all-purpose word that describes where we've gone astray: *duality.* Hence, our spiritual work to strive for *non-duality* is, rightfully, the critical endeavor of humanity.

I believe that my tiny contribution, touched upon by this book, lies in exploring some of those obvious, and some less obvious, polarities where we human beings get stuck.

This book contains both poetry and essays, paired with each other as you turn each page. Shouldn't it be either prose or poetry, or present a real or imaginal discussion? Indeed, that's the issue at hand. But let's take a closer look at the *two realities* to which I refer.

While studying for my MA degree in Transpersonal Psychology in the early 1980s, I encountered a teacher named Angeles Arrien. She was a cultural anthropologist, educator, award-winning author, consultant, and an inspirational teacher. She was the wisest of the wise.

I had the good fortune to encounter her in numerous ways as the following decades unfolded. In the 1990s we were both adjunct faculty members at Matthew Fox's Naropa University in Oakland, California, where we were paired as teaching partners. She was the primary teacher leading the morning seminar, which I sometimes attended.

One particular exercise of hers is quite central to a theme of this book. As a cultural anthropologist, Dr. Arriens studied numerous teachings from indigenous traditions throughout the world. One such group experiment was drawn from the Dogon tribe of West Africa, known to be a people with a profound connection to spirit and cosmos. She divided the class into groups of three students, each of whom would tell two stories, one *real* and one *imagined.* The partners of story teller would then attempt to ascertain which story was true and which was fictional.

At first glance this exercise seemed to be curious and unrelated to the world of psychology or mythology. But I soon got where she was leading us. In the Dogon tradition, those whose stories were most indistinguishable as to veracity or imagination were chosen to be the judges and wise elders of their communities. In our culture we might simply dismiss this talent as a clever parlor game skill, but in other cultures the relationship between so called reality and imagination is a highly treasured, and paradoxical, attribute of wisdom.

This book of essays and poems is fused with this teaching. We all appreciate the demarcation between poetry and prose, inventive storytelling and *the truth.* My writings attempt to revisit that literary frontier and expose it as yet another contradistinction that we tend to hold unnecessarily.

The relationship between prose and poetry, then, is more than simply two categories in the library. It reflects the two primary modalities of thinking and feeling that we tend to accept unconsciously as we study and learn. Most of what we consider to be education in our culture falls into the realm called prose. Even scientific charts and tables are essentially prose related and considered to be quite separate from poetry or other artistic forms of expression.

Modern physics holds the position that the physical world has a hard-wired limitation that keeps the mind from grasping both of those perspectives simultaneously. Think of the famous pictograph of the vase, that when viewed differently, is seen as two faces. It is impossible, they say, to view both of those realities at the same time. Quantum physics calls this phenomenon *complementarity.* I attempt to soften that hard edge between the worlds of mythos and logos and explore how they can walk together and view history and religion, light and dark, life, death, and god.

Science and religion, fact and Biblical text, are polarities that invite revisioning in ways that might actually have been more in balance in ancient cultures. I believe that religious fundamentalists have allowed themselves to be hoodwinked by scientific literalism (let's go to Iraq's fertile crescent and search for Noah's Ark). In reality, this muddled way of holding religious truth is more recent than we thought.

In my view, *complementarity* is the place beyond the physical world, where the arts of prose and poetry can dance as partners. Consider how dancers maintain their individuality and yet simultaneously merge into a holistic entity that transcends their separate components. In this spirit, the book presents short essays on the left side and poems on the right side of the pages of this work. At times the prose may have a poetic flavor; at times the poetry may lean toward the didactic. At times I hope that you might get confused.

In any case, I want to challenge you to re-evaluate your way of interfacing with long-held notions about science versus spirituality. There are mysterious forces at work and even more so, at play, in the dream-time reality of this ever-expanding cosmos.

Put another way: *What Happened to God, Religion, Science and Democracy?* is a direct function of the unique spiritual evolution of the western mind. I hope that these essays and poems will shed some informative light, and, some nutritive dark.

Chapter 1

The Unfolding Cosmos

The Cosmos Is Not Out There

As human beings we tend to see the small picture. We focus upon personal dramas, what's for lunch, this car, this job or even this career, issues that pale in importance to much greater things.

In some as yet unproven way, we are inextricably bound up with, well, the stars! Now this may sound like superstition or worse, but it may be much more significant than you realize.

Western philosophy and theology have separated humans from nature. We've objectified nature and made it a tool to be used and manipulated, and the consequences of all of this are not lost on you, I'm sure. But science and physics seem to be breaking down many of these questions about how the cosmos began, how it works, and how it relates to time, light and dark, and a whole slew of other mysteries.

If I may speculate here, the science or art of Astrology, ancient and modern, seems to traverse the terrain wherein the cosmos itself meets psyche. Even the ancient forms of our monotheist traditions used to acknowledge this reality. In Judaism for instance, *Mazel Tov* is Hebrew for good planets, or the wish for a favorable alignment of the planets. We've lost its true meaning and until recently, the possible scientific wisdom therein. The ancients may have known more about the true nature of things than we often suspect.

We Are Stars

You are not a tourist in this cosmos. So
enough with the guidebook. You'll miss the point.
You are a flurry of sparks of the big bang. You are a
flaming ember of the one hundred million-year chrysalis
fire-ball of creativity of the highest sort. You are a melody
line in the uni-verse, the one song that is singing itself and
we harmonize with it each moment that we laugh and cry,

smile and ache, dance and drum, snap and clap, move and
groove, so when we drop our masks and dare to chant and
sing, we start to strut our stuff with that sparkling cosmic
bling. So go ignite the mystical spark that invites you to
contemplate, meditate, penetrate, germinate and procreate.
Get out there

and realize
that the big bang is still bursting, that our cosmos is ever
expanding and we are its wide-eyed lusting partners, eager
to join the spiral dance. Dust off that mandolin, soften
your grumpy heart, buy a copy of Street Spirit, say hello to
friends and lovers, forgive your enemies, and honor your

precious teachers before you miss the point, that you are the
point: a shimmering, vibrating strand of light inside
a universe that continues its mysterious
hopscotch far beyond our
antiquated notions of
time.

Before Birth

Genesis is confusing at best. There are at least two versions of creation in it. Some of it is poetry, some of it is prose. You remember bits of it like darkness to light and making plants and animals before man and woman. A few thousand years after Genesis was composed came the Kabbalists. Somehow they deduced that in order for the world to be born a great contraction must have first occurred.

They called it *Tzim-Tzum*, followed by a great shattering called *Shevirah*, and finally *Tikkun*, repairing the world.

A few hundred years later (these days), physicists seem to ascribe to the latter description when observing the behavior of the cosmos. We've all heard of the big bang. Less famous is the oscillating model considered by Einstein, of successive universes beginning with big bangs and ending with *big crunches*, a great name for a candy bar by the way.

I admit that I'm a bit smitten with the contraction idea that seems to be in vogue in most of the models above.

The Great Contraction

Life doesn't begin with a bang; the
Kabbalists and the Quantum Physicists agree.

The *Ein-Sof,** and we it's cherished offspring
are beings of such dizzying and dazzling divine light
that we are first compelled to undertake
a great contraction of cosmic proportions.

The universe breathes in before it births itself.
And we were birthed by mother's contractions, so I suspect
that only after such cosmic shenanigans can we ever hope to utter
our first sound, take our first step, and rehearse our part
as we falter and stumble into our roles in the big dance.

So let's empty before we expand and forget all we know
and plunge with abandon into the still dark hollow
at the base of the great vessel where there is only the still point.

We can call her Lilith, Persephone, Kali or Mater.
She is always the dark mother who births all creatures
into a world where no things are separate,

where all actions emerge from contractions,

and where truly, everything dances.

* The endless reality.

Fire and Water:
(A Few of My Favorite Things)

Yes, I do think about fire and water a lot. More than earth and air. Somehow I got it that creation was more about a big fireball that burned for millions of years, or billions, and water then took over at some point, and "was upon the face of the deep."

When or how earth and air became visible eludes me. In the more mundane world where we live, it seems that fire and water are still the unruly ones, as elements go. Yes, we have earthquakes, but even those may be the result of some fire-related events deep within the core of the earth. My limited geological knowledge aside, I think about fire and water, write about them, make rituals around them, and at some subtle level I hold that they are the two seminal items that make up both cosmos and human consciousness.

When they do go out control, however, it is heartbreaking.

Fire and Water Unleashed

Wild fires rage and we notice
terror and destruction forgetting about
the spark that ignited the universe

and the cosmic fireball that burned
our galaxy into existence.

Waters rage, we see cities destroyed
and rail against such random cruelty
forgetting that we are water,

that we draw it from wells that sustain,
streams that enliven
and oceans that transport us.

We forget that tragedy burns in the bellies
of those living it, and yet, at such times
we're mysteriously joined to each other
in ways that normally elude us.

Fire and water mutiny,
we doubt the existence
of a loving and caring being
at the core of the universe,

forgetting that the very acts of
feeling, grieving,
arising and loving

are rays of light
emanating from the very source
whose existence we question.

During days such as these
we are surely forgiven,
but faith and doubt

are partners for life.

Things That Are Empty

I admit that I marvel at full moons more than new moons.

There is something about being empty that makes me quite uncomfortable. Intellectually I know that moons wax and wane, tides shift, people live and die and ... oh yeah ...

I guess that I just hit a nerve.

Emptying out was seldom a welcome teaching. Emptying the trash or having an empty cupboard or an empty wallet is generally not favorable. At times I fear that my life might also be empty but I also know that empty moments can be welcome meditations, and uncertain life transitions can actually be necessary and life giving. As babies we seem quite happy not being able to speak, walk, reason and a host of other activities that later seem so important.

This discussion is all about living and dying.

New Moon

So many poems about the full moon . . .
and so few about you, new moon.

The full moon is so Hollywood.
lovely bright and glorious,
overlooking a Zen master
on a still lake,

or surging through menacing clouds at
midnight in a vampire movie.

New moon, you have no P.R. advance team,
your image seldom looms over marriage proposals
or prom invitations.

But thank you for holding a place
for what is yet to be.
ripe with possibility and potency.

Thank you welcoming all
gods, wishes, and dreams.

Thank you for being
the un-ignited spark of creation
and the precursor to
each precious moment,

like a proud mother standing nearby,
barely seen,
but grateful
and content in
her selflessness.

About Now . . . and Then

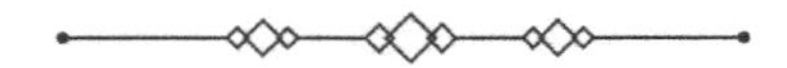

We all know about this concept, either through years of awareness of Zen Buddhism, various mystical paths, or Eckhart Tolle's important book, or just naturally being present to each moment, (this last being rare).

In reality, much of our inquiry into philosophy and theology are a mix of what happened, and its importance, with the *truth* of this moment being paramount. So how can this be? What importance do ancient texts hold about who did what?

Yes, it's a jumble for sure. I have no answer here but I will acknowledge the enormity of the issue and at least some of the pitfalls of religion's way of dealing with its ancient heroes.

The Power of Now and Then

Now
And then
are a spinning top
a whirling blur of what
may have occurred way back
when in Eden or Egypt or maybe Mt. Sinai
and yet from Buddha and Einstein and both Eckharts
we are learning that now and then are really one and the
same,
you no doubt remember how Moses' rod transforming into
a snake
and then soon parting the Red Sea was once the hottest
latest news
where those so called sacred texts contain moments where
we learn
that the stories living in our hearts are not meant to be
fact checked
but are instead the ever-burning bushes illuminating one
moment
that gets misconstrued as the retelling of a newspaper
story
and even worse, an ironclad prediction of a cataclysm
when actually the biblical narrative circles around
the great mandala sparks of universal truths
that are forever living and rocking in the
nestling arms of the great mysterious
author begetter and originator
of all beginnings middles
and endings of the one
forever unfolding
and amazing
Now

The Great Eclipse of 2017

You all no doubt remember where you were when the Great Eclipse occurred in 2017. It may have the stature of other momentous events: J.F.K.'s being killed, or the first moon landing, or 911, or the tragedy of 2016 and its aftermath.

I'm intrigued in so many ways about these sorts of events, especially the *natural* ones, those of the cosmic variety. The way we anticipate them and the lengths we go to in response are quite telling about human nature.

Central to this book is the intersection of cosmos and religion. Why, they might actually be twins.

Three Rebbes and The Great Eclipse

Three great Rebbes are standing in a field of groomed flowers
as the great eclipse approaches.

One inquisitive Rebbe asks his comrades if the reflective
moon and the radiant sun might be offering prayers to earthlings
and if so, what might be those prayers?

They confer.
The first Rebbe posits that a moment of darkness amid daylight
reveals the paradox of light and dark.

The second Rebbe affirms that the moon interrupting the sun
joins the kinetic forces of the universe with the reflective.

The third Rebbe speculates that the extraordinary alignment of sun,
moon, and earth are pointing toward *Shemayim,* the heavenly
residence of God and angels.

Standing amid uniformed rows of multi-colored dahlias and alluring
white tuberose, the Rebbes wait and watch as the miracle unfolds,
and all three begin to cheer.

In the field stands one perfect sunflower whose center boasts
tender black tentacles surrounded by pulsating red petals of dancing
light just as the moon caresses the sun, sending its
shimmering rays outward.

The eclipse approaches totality and the Rebbis say,
"Look how the sun and moon have joined, offering humanity
this special prayer."

What we and the Rebbes forget is that
the whirling planets with their heartfelt prayers

are always reaching for totality.

About Cosmos and Donuts

Once upon a time we thought creation was enacted in a flash.

Let there be light came after the darkness upon the deep, and the story proceeds from there. It's helpful to remember that the folks who wrote down that tale saw the world as, well, Israel.

No one then and there even imagined that China or Siberia or the rainforests of the Amazon might have existed.

Even at the time of Columbus, the world was believed to be flat and sat upon a turtle. Today only the right flank of a certain political party live in that mindset. In short, our take on the universe is constantly changing and it's difficult to track it all, much less appreciate the cosmic significance of it all. Notions of time and space and gravity were concrete and could be measured reliably. Now we know, at least some do, that we are being called upon to grapple with the notion that the universe is not static.

Things are changing, growing, expanding and contracting. So what does this mean for us? Should we be watchful and attentive to such theories or just stick to our reliable *guns*? Do our psyches need to shift along with this information?

I'm guessing you'll answer in the affirmative to these questions, and yet, at times, I need and want a simple time out. It's all flying by rather unceremoniously and causing my head to spin.

Space, Time and Donuts

Or is it time and space? They're both
getting arthritic and nearly bed-ridden.

Used to be, time could be trusted to be
straight forward like time lines, timetables,
even time shares. Linear and

dependable. No more, since Einstein
that's all up in the air—like space which is
altogether a different place. Space went

out in all directions, you could trust it to
be black and endless. No longer, even space
is curved or so they say, and beyond

the end of the cosmos might be nothing at
all. Consider that. Apparently everything
bends back in a circular fashion.

The cosmos resembles a chocolate donut,
round and covered with creamy darkness,
and sweet with nothing in the center.

Now these are uncertain times but this
chocolate donut sure looks good on this
cool October morning.

Chapter 2

Music & Sacred Sound: A Vibrating Universe

Drumming and Descent

Growing up in St. Louis, I viewed drumming purely as entertainment. It was about Rock n' Roll, top 40, and Jazz on T.V. variety shows.

There was barely a whisper about drumming as an ancient practice in a religious sense. To be a drummer was to emulate Gene Krupa and Buddy Rich, and possibly, be wealthy and famous. To drum was to reach *up* to God's heaven.

We didn't realize that music was audible spirit and that drums predated Egypt. Nor did we suspect that spirit could be found in earthly materials such as wood and animal skins. In fact, spirit seemed to exist only in a distant heavenly realm, far away from our daily reality. It took many years of living, schooling, and reflection to even begin to realize that a lower world existed. The ancient Greeks spoke about an underworld that was often conflated with the notion of a hell. Sorting takes decades or longer in a life committed to the higher mind. Western philosophy as we call it still struggles mightily with this split.

One thing of which I am certain is that drums take us to spirit's forgotten lower dimension.

Descent Into the Drum Time

This pulse inside my belly
echoes the throbbing that pours from my fingertips.

I'm watching the blur between skin and membrane
open an imaginal space where a moment ago
there was no space at all.

Silent waves within flesh and wood race
toward the round measured sounds of drum beats.

A collision

Light shatters

Walls crumble

And animals dance beyond the star-lit moonlight.

In the drum time we are riding waves
of rhythm, flesh, and curvaceous silken oak,
and—oh, feel the heat and moist rapture of that moment

when skin embraces skin, the ground trembles,
and we cascade headlong

into strange and steamy worlds.

Rhyming

As children, we assumed that poems must rhyme. We craved the rhythm and meter of nursery rhymes and Dr. Seuss books. Then one day, even after reading Robert Frost and William Blake, we were taken aback by Allen Ginsberg and also e.e. cummings, whose name he and we summarily stopped capitalizing.

It has now become poetically incorrect to use alliteration, rhyme, and various meters, not to mention the mixing of such devices.

But someone has to address this unfortunate dilemma, which I have attempted to do on the opposite page. It is my hope that you as reader will not summarily discount my poetry because it reminds you of Miss McGillicutty's fourth grade English lessons.

Orpheus was and is the Greek god of music and poetry. We separate those concepts, or at least distinguish between them, but the Greeks did not. I suspect that rhyme and other devices were so necessary in the oral tradition where stories like the Iliad and the Odyssey were memorized, and music and poetry were inseparable partners in that venture. I believe that it's time to reinvigorate modern poetic expression with music again. Relax and enjoy.

Poems That Rhyme

evoke an earlier time
when devices like meter and alliteration
seemed to be steadfast fixations
paying loving homage to poetry's Orphic foundation
with no ensuing crime
to the sublime

but among the many who have had the temerity
to craft poetic lines in modern times
concerning rhymes

there exists an undeniable enormity
of those who adhere to an obligatory conformity

to assuage the lofty expectations of listeners and readers
whose admonition to ascribe to what is now
proper erudition

precludes the time-honored convention
of using rhyme and meter in poetic recitation

but I remain undaunted in feeling the wondrous
heavenly rhythmic elation

of offering the world
the mellifluous expression

of the one great
universal vibration

Music and Nursery Rhymes

Concerning rhythm, nursery rhymes are right out front about it. *This is all about rhythm,* they seem to say without obfuscating the subject.

The children need to get to sleep for their health and well-being, and, we parents are desperate to have a moment to ourselves. Whatever gets the job done is what is required and for millennia, the gentle vocal swaying of *Rock-a-bye Baby, The Itsy Bitsy Spider*, or Dr. Seuss's miraculous tome *The Cat in the Hat* have soothed and entranced both children and adults.

Often we put such contrivances into the category of non-literature. Yes, they are clever and rhythmic but some say those simple melodies and rhythms can't be taken too seriously as *music.*

The rhythm of these simple ditties penetrate the core of who we are. Quantum science now speaks of vibration, waves and particles, and alternating realities in a manner that screams . . . rhythm.

From Pythagoras 2500 years ago, to the French physicist Louis de Broglie in our lifetime, the wave nature of matter and electrons seems to suggest that humans, animals, plants and minerals are all pulsing to some similar and great essential truth. Call it religion or science. Both seem hungry to explore the nature of what we call the universe: the uni-verse, the one song.

Shabbat Shalom Went Up the Hill

For a moment those words roll mellifluously off the tongue before that pesky pound of left leaning brain mass alerts the system that a foul has been committed. Then, with a whimsical smile a yellow flag is yanked from the back pocket of the referee.

Years ago, our three-year-old daughter merged those two worlds, oblivious to the inherent wisdom and the delicious rhythmic flow of her creation.
Shabbat Shalom went up the hill!

The Sabbath greeting of choice for Jews wedded to a nursery rhyme, suspended in stop-time, like exultant tango partners.
Who leads and who follows is in question
like Ginger Rogers and Fred Astaire.
As we know, Fred always got the top billing as does:

Judaism over the eyes closed, reverential chanting of the Sh'ma; Catholicism over the imitative counterpoint of Ave Maria; and Buddhism over the rhythmic murmurations of the Lotus Sutra.

Religion may have been birthed at Olduvai Gorge as a dispirited clan marveled at Lucy's bones, 60,000 years before the modern world found her remains, or it may have been the first song whistling its rhapsodic joy and grief across that East African savannah.

As for our lifetime: there was Coltrane, who entwined music and a nameless spirituality into *A Love Supreme*—perhaps the title of the next new religion?

Misunderstanding Drummers

I often teach and talk about drumming and spirituality at churches, temples and universities, and begin with a story about the movie *City of Angels* (Meg Ryan and Nicolas Cage) where Ryan, a doctor, falls in love with Cage, a heavenly angel in LA who meanders through hallways and streets with a strange look in his eyes. He appears often and mysteriously in numerous contexts in her life. Fearing she is either being stalked or has fallen for the wrong guy, she finally confronts him and asks three questions.

"Are you married?"

"Are you homeless?"

"Are you a drummer?"

Need I say more? When I first saw the film, I was the one in the theatre laughing loudly.

The culture at large is suspicious of drummers; fellow musicians joke about them, European traditios demean drummers for being other, outside of the norm and representative of native cultures, *the dark continent*, and particularly in America, people of questionable repute and lifestyle.

Trombone players or violinists are not held with the same sort of suspicion. Something is afoot here that requires examination. Is it about culture or sound or spiritual inclination? Perhaps all of the above? I invite you to read on and make your own determination about this much maligned and misunderstood phenomenon about drummers.

Drummers!

Unbeknownst to biblical scholars, behind the tree of life in
the Garden of Eden stood drummers. Now bear with this
outrageous claim for a moment and consider the notion that

after Adam and Eve grappled with the apple of evil and good,
it was drummers who chopped the sacred wood and hollowed
out the first drum thereby soothing that theological schism

with the great-log-a-rhythm that still accompanies the sun to
rise and the late afternoon shadows to drift toward evening's
obsidian crypts. Yes, it was drummers who set the tempo

for Joshua's trumpets with primordial djembes and congas
as Jericho trembled and tumbled while angels mumbled such
shock and dismay at this raucous display of rhythmic play.

Yes it is drummers, misconceived by critics with their
fruitless analytics who are better perceived by kids at
play than are those doughty violinists who just concluded
their triumphant armadas of sonatas.

Yes, drummers, whose weathered hands dance their dream
drumming riffs on wood and skin, are back in the temples
and altars for moments where spirits needs raising

and angels need praising to spark the movin'
and a groovin', the rockin' and a reelin', and the cosmic function
at the junction, where we sanctify with holy wine and sacred bread

the Rumba, Jazz, Samba, and Salsa, Flamenco, Fandango and
the sensuous Tango, reminding us that we swing in a universe
that pulses and palpitates the yearning heart
with the one vibration,

be it final ending or primal start.

Silence and Sound

Many of the great heroes of the world's great stories were musicians. How could that not be so? The big bang is happening: not the great fragrance or the great culinary taste-off.

We see artistic renderings of Jesus and Buddha surrounded by animals who are entranced by the auditory vibrations of the world's most prominent spiritual teachers. None is more famous in this regard than the Greek, Orpheus, who has inspired innumerable tales and inspired many sound merchants, such as myself, to find the link between music and human feelings and the vibratory core of being itself.

Music is the primary tool of the shaman, the healer, the priest, the minister, the rabbi, and all who explore the soul.

At the core of all music lies eternal silence. Eternal suggests something out of the confines of time, so, like Orpheus, we must traverse the boundaries of this world and learn to negotiate other worlds as well.

Listening

Listen inside silence
for the seed of music not of
this world.

Orpheus plays music for the gods
in the underworld.
Grief is his conductor.
Joy has been spotted
briefly,

They rendezvous
for the last encore.

Heaven joins the still earth,
as all beings that truly listen
within the silence
yearn to be released.

Up or Down the Big River?

Take a moment and picture the image you have of the Mississippi River. Do you see it flowing downwards (from Lake Itasca) in Minnesota or upwards from New Orleans? There is no right answer. But for me the difference is quite significant.

There are the laws of gravity, our prejudice toward north signifying up, and south signifying down, on our maps. But consider history and slavery, culture and migration, and yes, music! For me and others, what happened with rhythm, music, dance, and celebration far outweighs the laws of physics or cartography. For me, the world spins on an axis of music and dance and the people who created it and continue to transform it.

A history of the world could rightfully be the creation and transmigration of the variety of musical experience. (Note William James tour d' force: *The Varieties of Religious Experience.*) Religion and music are twins, maybe identical twins.

Up the Mississippi

It went up, starting in West Africa
the rhythm, carried by the tortured
and dispossessed, still managing to
possess the best despite the
whips and slave ships.

They couldn't stop it.

Up the river it sailed on backs
that sweated and bled,
From New Orleans, Memphis, and St. Louis,
it spread and bred:
Louis and Ella, Billie and Miles, Duke and Trane,
and James Brown: *who's gonna'*
drive you insane.

Tip your hats to the nameless
who tugged us all up rhythm's river
and made us reel and rock,
cause for heaven's sake, our stay on earth
includes a
tic and a tock.

Evolving Language and Orpheus

Sometimes great poems have to be re-translated, even after highly talented and world-renowned authors have set the established mark of excellence upon these works. Language and nuance change dramatically over a century, and at times, even within a decade.

The story of Orpheus has been a guiding light for both my inner and outer lives for most of my adult life. I've searched for translations of Rilke's poems that spoke to me, with disappointment, even confusion. Here are some examples of translated lines written many decades ago from various authors (I've disguised some of the lines for obvious reasons).

"only the one who lauds without cease"

"dissolves many asunder"

"may present the intimations of endless praise"

Forgive me for displaying these out of context, but we've all experienced the awkwardness of *classic poetry* and have perhaps been refreshed by the emergence of modern poetry over the last century or so. One of the issues is simply the words and phrases that were then in contemporary vogue, that seem arcane in today's world.

Many highly educated folks I know avoid poetry because it can seem so abstruse.

In that light then, I've tried my own *up to date* translation that no doubt, will irritate someone in ten or fifteen years from now, who will create his or her own translation with words, phrases and metaphors that reflect the zeitgeist of their time.

I welcome them to quote and critique me. I would, however, ask you, the reader, to wait at least a week or two before commencing with this undertaking.

Sonnet to Orpheus #9 (Rilke)

Only those who play the Orphic Lyre
in the world of the departed will fully
know the rapture of blessing one's days.

Only those who have eaten poppies
with the dead will revel in sound, in
touch and gaze.

Though the reflection in the pool dissolves
before your eyes, the essence of things also
lives in places dark and nocturnal.

Live life in both worlds, where voices are
enchanting and the heart of music is
eternal.

(translation: Bruce Silverman)

Chapter 3

Light & Dark & All That Swirls Around Them

The Dark That Is Beyond Dark

It seems that we are so confused about light and dark that the first tutorial required is to acknowledge that beyond the normal form of light (from light bulbs, let's say), there are more subtle and esoteric forms of light. It's the same with darkness. Something more profound is lurking behind the dark evening sky than what we take for granted.

This land beyond the normal is where metaphor collides with cosmos.

Exploring light and dark is not simply a literary event; it's a swan dive into the cosmos itself. The deep pool to which I refer is, I believe, the main event of the planet and beyond. Genesis got that right at least.

If we as a species are to take the next step forward, and if there are any steps left that are not crumbling, then it will be to balance what science calls *truth* with the great cosmic unfolding often called *creation*. (Don't confuse creation with *Creationism* please; it's just another *ism*. And it's at least as toxic.)

Loving the Dark Mother

We hear about the dark night of the soul, dark humor and

the dark days ahead , darkness always seems to get a bad rap.

There seems to be no end concerning our aversion toward it.

But I love the darkness for hovering over the face of the deep.

I love the darkness for showing me the brilliance of the stars.

I love the darkness for holding me as I fall into
the land of dreams.

I love the darkness for pointing to all that is unfinished in me.

The light we radiate has a mother named darkness that resides

in a still place inside us and all around us.

To drive her away is to

reject one half of creation.

Be nice to your mother.

We humans have some deconstructing to do about some basic realities of the universe. This chapter explores light and dark.

Is there a more misunderstood concept?

Sometimes it's wise to look beyond the everyday aphorisms about light and dark that abound. Go back a generation or two, if you dare, and pull out some family photos. You just might discover a simple but powerful teaching if you look carefully.

Be sure to bring along a special kind of lens for this assignment.

Black and White Photographs

In the basement I rummage through
tattered boxes stuffed with precious black
and white photographs of parents and ancestors.

I discover Aunt Minnie at age twenty-five,
with jet-black hair, dressed to the teeth,
alluring, with garlands in her hair,
sitting alone in the dance hall.

I picture
someone missing his chance to
dance with this dark beauty, alone
and waiting.

Too often we hesitate to walk across
the worn oaken dance floor and reach for the
hand of the neglected wall-flowers of our lives,

dressed in dark uncertainty, but whose
titillating tangos, nonetheless, may be
calling to us from the corner
of the ballroom.

Too often we hide in the familiar bright light.
Like those old photographs,
our souls require darkness
in order to develop.

Things I Learned in Sunday School

Speaking about darkness: our deep-seated fears. I still see my third grade Sunday school bible book, Adam sitting, broken and cowered as evening fell and the cloak of night portended endless darkness, snuffing out the spark of creation after one blessed day.

Darkness had designated and stamped the universe as an unfriendly place.

Some relief for Adam (and me) came in the rendering on the next page as he awakened in morning's sunlight. But the damage was done. Darkness was, henceforth and forever, scary.

Let's Give It Up for the Dark

Each Holiday season there's so much
buzz about light rescuing us from the
dark. We treat it like a convicted felon.

Out come the Hanukah candles, the
advent calendar, the blow-up Santa
and the colorful illuminated reindeer,
all of this to drive away the dark.

The source of this hullabaloo is
the winter solstice, tracked by
astronomers with barometers,
astrologers and mythologers
who now inform us that
the universe is:
68% dark energy,
27% dark matter,
and a mere 5% light.

(I don't quite get it either)

If these experts would re-read
Genesis they'd remember that
darkness was and still is upon
the face of the deep and is not
simply an affliction but a
cosmic reality, fertile,
pregnant and generative.

The brightest light emerges from
the still darkness of the cosmic
womb.
So now that religion and science
are on a collision course, let's
dare to befriend the darkness.

It seems to be gaining on us anyway.

The Glow Behind The Things You Know

There's light and then, there's light. Over a life-time I've learned that young children see light in magical ways that busy adults do not. Also, those adults miss the awesome experiences that mystics and seekers often describe in moments when the ordinary light of day is displaced by an altogether transformative sort of light that we tend to see only in medieval paintings and occasional sunsets.

Seeing That "Other Light"

Have there always been those gentle ethereal
lights hovering around the flames of candles?

Look carefully and see their glowing gowns,
dressing and caressing the fiery tops,
bowing slightly as if to some nameless god.

Even the street lamps send their hypnotic silver
swords outward, illuminating the pearls of rain
as if to announce the entrance of some eagerly
awaited queen of the night.

Softening your gaze you notice the light around
the bodies of fellow humans, sensing that this
harness of gold connects you to all beings.

So much light in front of your eyes.
Flames illuminating gods and queens,

and
that person sitting next to you on the bus.

Not Knowing Where the Hell Your Ship Is Going

Light and Dark are polarities but like all polarities they are not mutually exclusive. We tend to think of light as visible but that's not always true. Even darkness, that which we assumes lacks light, can be seen and felt.

Prophets and mystics have long talked about such matters and metaphor seems to be the best way to get at this tricky business. The Sufi mystic poet Hafez (CE 1325-1390) used a ship as the vehicle to describe what we can know and not know. Like other great theologians and spiritual teachers, he suggested that one has to jump ship and leave behind all that might be familiar, even or especially the notions of god and religion.

.

You've Jumped Ship, What then Hafez?

The great Sufi poet, Hafez, said:

"The Great Religions are the ships
and poetry the life boats.
Every sane man I've ever known
has jumped overboard..."

But you may ask: *what then Hafez*?

Then we'll drift across uncharted seas in lifeboats without the antiquated provisions of clerics.

We'll survive by drinking holy rain water, catching luminescent spirit fish, and making midnight prayers of the heart.

Then after years or decades, we'll board another great ship that leads us onward, but not back to the familiar ocean of certainty.

We'll sit and join hands with those huddled in the dark recesses of the ship's steerage, who have dared to leave home forever in search of unknown lands.

Gardening

Gardening subsumes all things. Since the universe is expanding and all things are changing, growing and withering, gardening seems to cover it all. Mr. Rogers certainly taught that maxim daily, and you might recall the book, *Being There,* written by Jerzy Kosinski, and the movie of the same name starring Pete Sellers, the *simple minded* gardener who sees the world only through this prism.

Kosinski was pointing to a truth that swirls around us each day and each moment. At times we gather to create rituals that remind us of what we already know. We humans need constant reminders. The character Chauncey Gardener knew how to lie this simple truth. It's easier for fictional heroes.

The Garden Room

A dozen men arrive in a room within the walls of a
a small Jewish community complex leaving both
religion and politics at the door.

The circle features soft drumming, candles,
Rumi and Hafez, Blake and Rilke, and
some age-old circle magic.

In the halls nearby the fourth night of Hanukah is
swirling outside the Garden Room with food,
art displays, music and celebration.
Both inside and out of the room:
waltzing with light and dark.

No, no strings of colored lights or giant
menorahs, only darkness as centerpiece, punctuated
by haphazard candles displaying the shadow forms of
each man upon the surrounding Tuscan Yellow walls.

The cold winter
evening goes about its cycles under an empty moon.

This night, both outside and inside gardens expose the
roots of living things growing, being lovingly tended by
the spade of soft drumming, the hoe of silence, the seeds
of exploration scattered randomly in a cave whose agenda,
is unscripted. With hands and hearts feeling their way along
narrow passageways, carved by the stories humans share,
men are transported downward
through the labyrinths that both enlighten and endarken.

This night, with Hanukah blazing steps from the door,

some are attending to the shadows
that are forever cast upon the numinous cave walls,
often unseen, but patiently waiting.

Still Cursing the Darkness?

The tree of life in the garden is about *the knowledge of good and evil*. Maybe we used to know, like the gnostics who felt it in their *kishkes,* (Yiddish for one's gut), but knowing today pretends that it can keep away the curse of darkness. Darkness is often however, simply not knowing.

The most common strategy for dealing with darkness, sadly, is to make what is uncertain or unfamiliar, evil. A lot of that seems to be going around these days.

You've heard the cliché about lighting one little candle rather than cursing the darkness. I think we need to dig a little deeper.

The Curse of Darkness?

These are dark days
we often say.

But darkness holds us
in our not knowing.

It's expectant and
births light.

The enemy . . .

lurks elsewhere.

Chapter 4

Where Brains, Minds, & Dreams Collide

Beyond Brains

I have friends who study brains: how they are constructed, how they grow, and of course, foremost in this world, the aberrations and pathologies that abound, and the medicines and practices that manage or even cure these maladies.

As I have studied and taught and continue learning, increasingly I encounter the word *mind.* It seems that brain and mind used to be interchangeable concepts, but for me and many others, no longer.

Curiously, the more brain science emerges as a phenomenon in our scientific and psychological literature, the less stature the organ of the brain seems to have and the more primacy the word *mind* seems to have.

Something quite important seems to be happening that has, well, religious, or at least spiritual, implications, and I believe we must pay close attention to it.

I also believe that we need tools like poetry and comparative mythology to adequately describe what's transpiring.

It's A No Brainer

Once upon a time, as in a dream,
there were giant gilded birds
that built their nests in the
branches of cinnamon trees.

We humans flew on the backs
of those great winged creatures,
marveling at the verdant mountains
below and breathing in each precious
moment in a boundless spirit world.

Then over time our brains grew,
we were taught to *know thyself,*
to think and therefore to be,
and that *we are what we eat.*

Great thinkers now assure us
that the winged creatures of
yesteryear were but brain
secretions.
So the question arises,
by even greater thinkers:

Does all consciousness reside
inside the scope of the human brain,
or does it fly on the wings of birds?

To me it's a no brainer.

Miracles in the Strangest Places

Where do miracles happen? Crossing great seas? High on mountain tops? They occur in religious texts for sure but what about these days?

W.B. Yeats had a transformative experience in a London shop over a century ago where for no apparent reason his body "blazed." This was no fever; it was not an experience simply of the body. I sense here a bit of a burning bush in his this poem of amazement.

At times, little miracles can unfold in strange places and at least for a day, allow one to see with new eyes, open one's heart and make the ordinary quite extraordinary.

Two Miracles at the Café

At the café while sipping the froth of my cappuccino,
I dreamed I was seated at a great table immersed in a
devotional discussion about Joseph the great biblical
dreamer in the death pit, betrayed by his brothers, sold
to a passing caravan where he became
a slave, a prisoner, a dream interpreter,
Pharoah's right hand vizier,
and one of the saviors of his people.

At that great table, stillness suddenly reigned, greeting
the arrival of Joseph bearing a dish brimming with the
aroma of manakeesh, cinnamon and nutmeg, while our
spoons, filled with light, fed ourselves, our neighbors
and strangers.

Gazing inward and upward, we saw angels tittering and
grateful, nodding their approval of our attention to the
miracle unfolding before our awestruck eyes.

Back in the café, there I sat, near a beaming mother
cradling her infant, perfectly adorned with what appeared
to be an Egyptian prayer cap now the object of most eyes
and moist eyes.

There we were, torn between the holiday canticle playing,
the morning coffee and the infant Moses-baby Jesus and
the universal child of the world, cuddling in the dark corner.

For a moment we all gasped and grasped the secret of the
omnipresent miracle before us as we were invited to wake
up, smell the coffee, and cinnamon, and pay attention to
the palpable joy bursting forth from mother and child
in the secret corner, now an open secret,
at the café.

Where Do You Spend Your Time?

We tend to think we make significant choices each day: what to eat for breakfast, who will earn our vote, who will be our friend, or what career moves we might soon make. And so on.

What is less obvious is where we choose to spend our time with our mind. Here I speak of mind vs. *Mind.* We often follow mind, follow to do lists, and jump from issue to agenda with nary a thought.

At times it's wise to stop and review what we've been doing unconsciously, by coming into the present moment where we notice the thoughts that fly in and out, and yes, even the thoughts that might fly in next.

This might be the most important task we accomplish in the course of each day. I wish I would attend to it more often. How about you?

Inside Your Mind

On your daily walk you spent a bunch of time
inside your mind when you could have
breathed in the fragrant spring air
abloom with tangy mustard flowers
and their herbal brethren of a kind.

Years ago at the spring dance
you ruminated about whether you
had even a remote chance,
and thereby missed that fetching
and life altering sidelong glance.

Recently, sitting on your tattered divan,
your brain concocted an elaborate plan
that wouldn't pan,

While out your front door
there lives a world of sorrow,
waiting for you to ease
a frightening and tenuous tomorrow,

Awaiting a simple how ya' doin'
to rescue you and the world
from the mental pot
in which we're stewin.'

In fact, you and I can be so lost
in our head
that the dancing bees
and the swaying trees
might think we are dead.

So instead
put your tired mind at rest
and count the ways
that both you and I
are totally
blessed.

Illness as Teacher

I'm not going to speak to you about the flu per se. What interests me more is the dilemma of living, experiencing pain, illness, old age, and death. At a certain stage of life, this discussion takes on more resonance. I am in that stage.

Am I, then, morbid or at least distracted from the joy of living?

Am I depressed? Am I missing the essence of the gift bestowed upon me by being an aging senior?

I don't think so. I have always been intrigued by the subject of death and what some call the underworld. Be it Thomas Mann's *The Magic Mountain* or Joseph Campbell's various tomes or Haitian death rhythms played on ceremonial drums, I have gravitated to such subjects and life styles for my entire adult life. In the language of the wise Aboriginal Australians, I spend a good deal of time ... in the *dreamtime*.

Though we don't always hold it in this fashion, religion derives from the Latin word *ligare* meaning to bind. Now we've entered the dangerous terrain where words and concepts fall short.

I hope to bind more moments of my life to the awareness that I have those moments in the first place. I want to slow down, experience more of the beauty of silence, and meditate both sitting and walking in nature. Call it what you like: mindfulness, gratitude, the witness, or even that word, *religion*.

The Flu, Reconsidered

It feels so good
after having the flu.
Breathing easier the daylight seems
brighter, life seems more livable and
each moment more precious.
With this secret we can
sometimes scold ourselves
for those times when we're
stuck in drab everyday consciousness.

But must we catch the flu to remember
this clandestine truth? Or worse, must
we experience the horror of a loved one
in distress or in danger before we truly
relish the miracle of their presence?

Must we bargain with gods and angels
before we can savor the multitude of
blessings that envelope us? This food,
this friend, this life, this breath?

Gratitude is the divinely sculpted stone
path upon which we walk. All too often
we tend to the trivial weeds
of suffering
instead.

Giving thanks is the mystery within the
inner sanctum of the temple. Celebrating
how blessed we are is essential.

Having to first get the flu is optional.

Imagining My Miwok Neighbor

Sadly, I have no Miwok neighbors. Their whole culture seems to have been obliterated by the Eurocentric culture of which I am a part. In that sense, I can only imagine having a conversation, much less a friendly relationship with a native of the area where I reside in northern California.

The Northern Sierra Miwok people (as they are now called) lived here for 4-6 thousand years, as far south as Yosemite and as far north as Mt. Diablo, whose peak I see from my bedroom window.

Only in the last twenty years or so have most of us thought to honor these people in our own sacred gatherings, be they secular or religious. This oversight is heart breaking and inexcusable. The people who live and thrive on any land for this amount of time, or far less, not only deserve better, but are a part of our collective psyche. For my unconsciousness around this tragedy, I humbly apologize.

My Miwok Neighbor

Some days the walking path just steps from our door
is a place of bucolic wonder; it accompanies the canal
that meanders gently alongside. Nearby is a working farm
selling fresh vegetables. I also recall their neighborhood
hayrides that filled our children with such joy.

I revel in the display of the mustard
flowers and orange poppies abounding on the
path where playful squirrels watch the wild geese
and mallards make their pit stops flying north or
south. For moments I'm alone in this touch of wild.

Some days I dream of Yosemite's daunting cliffs, or
the storied Redwoods, but where I stand has its magic.
Just then neighbors greet Annie, our three-legged dog,
under the green foliage arches as bicycles fly by the
Live Oak and Modesto Elms hovering nearby.

On this day a nagging question arises: What long deceased
Miwok 'neighbor' might have walked on this land two
hundred years ago?

I imagine his ghost whispering to me as he views,
in horror,
the encroaching civilization around this tiny island strip of
nature and just maybe, his approval
of my fervent gratitude

for the beauty and serenity of what remains.

Those Crazy Lunatics

For most of my life I have felt out of the mainstream, not just the culture at large by the way. Even as I set my career in motion in my early thirties, I seemed to have always been a bit out of the mainstream of the non-mainstream. As a drummer, I gravitated toward working with modern dancers and playing rhythms that merged with movement, the way African and East Indian cultures do. As a therapist, I chose the transpersonal (a.k.a.) spiritual dimension of the craft, rather than the more traditional or *acceptable* wing of psychology. Even the mythopoetic men's group I facilitate seems to be unlike any other I've encountered.

Few words adequately described the people and cultures that have enveloped me over the last five decades. Luckily, I recently encountered a sentence from the writings of Jack Kerouac that used an exciting term:

Zen Lunatics. Eureka!

Zen Lunatics

(a term coined by Jack Kerouac)

Even in 1954 Kerouac Jack had the knack of knowing that a
spirited Zen pack would one day emerge and finally tear open

that star-spangled puritanical gunnysack strangling
the sleeping American Zodiac.

It's clearly our calling through outrageous and vivacious
acts to bring down those heat-seeking missile epistles
that deny all who display any figment of
dark pigment, a face too tannish or an accent
too Spanish.

Yes, I've had the good fortune to hang with such
a gang of jacks, of kings with spades, and clubs
that transform into talking sticks for Zen lunatics,
with bright diamond minds and open hearts,

that make an end run around the ten-ton anchor
of our civil rancor and then fly into an end zone

far beyond what's known, or owned, or cloned,
toward a different way, where there exists
a glimmering gateway of genius and justice,
adorned by crimson roses,
a wide welcoming gateway,
that never closes.

Chapter 5

Notions of God

The Precise History of Religion for the Last 10,000 years

Can there be a subject trickier or even more dangerous than the name of God? Human beings have been murdering each other over this dilemma for thousands of years. So it is with great trepidation that I approach this delicate endeavor, with both prose and poetry. I recount the last 4,000 or so years in the poem opposite this page.

Clearly the task I'm undertaking involves hubris and reductionism. It is helpful to know a bit of cross-cultural anthropology, ancient mythology, migration patterns of peoples throughout history, and, most of all, to have some compassion and roll with the huge leaps and bounds I make synthesizing millennia of complex events into short irreverent bursts.

Hiding inside these ruminations is an element that is often overlooked while discussing religion. I believe there is a psycho-spiritual wheel of development, much like the stages of ego development we studied in our Psych 101 classes. It is not directly addressed in the poem to follow but I invite you to consider its efficacy none-the-less. Suffice it to say that the developmental component of religious practice is inextricably bound to psychological development.

Many friends, teachers and mentors I have most respected in this life have indeed morphed their religious practices over the course of their lives. Some have stayed within the Jewish and Christian traditions so dominant where I grew up, and to that I would add the scientists and agnostics as well. But the great Christian mystic Meister Eckhart described the process of "leaving God for God." Here he is encouraging each of us to leave the imprinted notion of god and religion that was first thrust upon us in order to pursue whatever notion of god or religion might speak to the deepest essence of who we are. For me it comes down to emptying out before we can fill up.

It scares us when we dare to leave behind the religious teachings that no longer fit for us, and yet, it is incumbent upon us to do so.

In The God Kitchen

Out of the Arabian Desert four thousand years ago came an unknown monolithic, jealous, and all-powerful male deity. We embraced him, grouchy moods and violent temper, for thirty centuries till Joseph Campbell and Starhawk informed us that he had really been a she for for the previous six thousand years. Today's door-knocking evangelists, however, never got the message; they are still selling that Sistine Patriarch with a theology as black and white as the suits they wear but far more threadbare.

More recently some of us sought the gentler version of divinity, the *Shekhina** to temper *his* distemper. We tried varieties of the all-nurturing earth mother, soup ladle in hand, until we read the fine print: "Warning: this product contains no MSG but is made in kitchens by Kali and Pele with traces of fire, skulls, and dragons that may cause severe spiritual dyspepsia."

What we longed to stuff into a grocery bag of religious certainty and monotheism has more recently spilled all over the God kitchen into finger-lickin' splats of neopaganism, steadfast drips of Kosher-style Buddhism, frozen chunks of scientific agnosticism, and a host of other concoctions.

But there is no cookbook that captures the savory taste of the divine.
Words make tired recipes, cannot describe this sumptuous feast,
and will become fermented dangling participles.

My advice is to abandon the crusty old word pantry and head for the nearest desert wilderness where you will be seized by a presence that is eternally becoming. There you will encounter glorious displays of deep purple dreams and revelations, for as earth is now Gaia and sky is now Cosmos, we cannot keep flexing those bronze-age brain muscles by giving *you know who* a gender, a proper name, and a Facebook profile.

It's time to drink the potion that will rid us of the notion of defining what sparked the fires of creation and fashioned the depths of the ocean.

*In Jewish mysticism the feminine component of God.

How You Toss Around the Word God

We live in a world that purports to be rational. We actually have far fewer religious holidays (days off you might note) than did medieval Europeans.

Science has grown up a bit even though the right wing of the Republican Party pretends that it no longer believes in science as it did fifteen years ago.

(It became financially inconvenient).

We toss around the word god constantly, however, and do so carelessly. *God damn it* comes out of the mouths of babes and elders alike. *God help me* sounds like an earnest theological plea but is really just an exclamation of sorts. In a profane world the use of the word god is like an all-purpose wrench that we keep in our back pocket, a tool that can be used for a variety of tasks when we need a handyman version of dynamic language.

How much time do you actually spend invoking the divine force of the cosmos without ever realizing what you are doing?

Half the Time

Half the time you speak the word *god* carelessly,
not really considering the theological implications.

Half the time your doubts are so strong, you don't speak.
You don't really believe what you'd like to believe.

Half the time you weigh the evidence and feel hopeful.
After all, you've witnessed what can only be an act of god.

Half the time you are embarrassed to even broach the idea.
Your friends are far too sophisticated for such nonsense.

Half the time this whole business seems like a waste of time.
Why are you spinning your wheels this way?

Come to think of it, you've now spent five halves of your time.
How did you pull off that trick?

An act of god.

The Wretched Mix of God and Gender

God and gender are a confusing mix. Humans personify the divine and have done so for eons. Who really knows what the first humans did? But we know about historic Goddess religions and then monotheism—and now?

I believe the greater issue is image and holistic perceiving.

We humans have the capacity to grasp the whole, but binary thinking and separation seem to be the order of the day, and have been for thousands of years. Perhaps that is what being human is about, some kind of exam to see if we can remember what it's like to be whole and loving, and that we are truly one.

This would require a daily practice of attention to the whole: called mindfulness.

God Does Not Have Genitals

God simply does not have genitals, and this truth
would seem to be
obvious, yet we took it for granted that he was
masculine and now
we surely know better. Like it or not this gender issue
is seared into
our visual brain boxes just like Coca-Cola and Kraft
Cheese singles.
But how did this confusion about gender
and god actually begin? I can only guess
that we stopped gazing at the evening sky
or stopped growing food in little gardens
or invented the mixed blessing of writing.

Now we can continue to list the ways that we have
plunked what is invisible ineffable and
incomprehensible into little boxes
or we can simply walk outside
right now and marvel at
the Milky
Way.

About Jesus and Jews (not Jews for Jesus)

It's a sticky wicket to mix the Jewish experience for the last two centuries with the story of Jesus and the Christ (the anointed one). What was a very Jewish prophesy about the coming of a Messiah erupted into a world-changing phenomenon that I wouldn't dare to characterize here.

Suffice it to say that I've been confused about it all my life. My parents told the following true story that drives home the point. Our reform Jewish family in the fifties celebrated both Hanukah (our religious observance) and Christmas morning present-giving, Santa Claus and all. This apparent sacrilege does indeed lead to the essence of the story.

One such year my brother and I were stationed outside our parents' bedroom door at 6:30 AM waiting for the moment that we'd all gather.

Said Ken to me: "Bruce, do you believe in Jesus?" (Now imagine my parents' horror hearing this question, wondering if, in their desire to assimilate, they had gone too far and committed some irreparable harm to our religious souls). There was a long pause as they awaited my answer which was: "I don't know, what's his last name." Somehow we were all off the hook; both parents and children could and did laugh about this event for decades, but the confusion about how to hold Jesus in our minds and hearts to this day confounds most Jews. In general I'd say we haven't made much progress.

Oh Boy! Here It Comes, a Poem About Jesus

The shock. How did it happen?
It wasn't supposed to go like this.
You see, it wasn't easy in 3 BC
(we all have to call it that now)
We were all waiting for the messiah as the
next king of Israel would bridge the gap
between spirit and politics and along came
the radical rabbi into that cauldron of
cruelty in Judea and everyone knows
how the story goes.

Our Christian brethren get a free pass—
if you've behaved you're saved.
But for the Hebrew clan there's a
quite different narrative in store.
Jesus becomes a looming shadow.
He may never waver from granting
some a divine favor and, he's the ever kindly
but ominous elephant in every room whose
presence is felt from cradle to tomb.

As Jews we choose not to embrace his divinity
or Mary's virginity, but
he's become their Chief Rabbi, sitting on high,
in that place where we just cannot dare to tread
and the tired posturing from both sides of this
this theological divide distracts us from the
only real kernel of practice that we can truly
embrace before we are dead, so please
take my hand

and we will partake of this one and only
divine moment instead.

And for the sake of all
let's put this discussion to bed.

Wondering Who's Running the Show

Ostensibly we live in an enlightened age. Nietzsche's metaphor about god being dead was a historical observation, not any sort of a metaphysical creed. Guardians of *psyche* are more likely to be found in therapy offices than in churches or synagogues. *Psyche* means *soul,* not mind, but that notion might confuse a lot of people these days, even professionals.

All of this begs the question as to who or what is of greatest import? In other words, *who's in charge here?*

Who's in Charge Here?

I have doorways inside of me
lodged between fear and risk,
between doubt and faith,

Too often I can't break through
their steel girders.

Sometimes a violent storm
blasts through me and I
wonder if I am being
tossed willy nilly by
some malevolent
being.

I hide in moribund safety
and then, in time, some spark
ignites a daring new vision.

There's a rhythm to all of this and
a curious harmony that baffles me.
But ultimately

the conductor seems
to know what he's doing.

Chapter 6

Themes and Currents of The Days of Awe

Prepping the Vehicle of the Self?

Before the High Holy Days is the month of Elul when Jews begin the process of repentance and other practices that are a tune up of sorts. The intensity of the High Holy Days can emerge suddenly if one is not *prepared.*

Looking inside is a meticulous and labor intensive project.

Elul

In this month we are called to do maintenance and care for the chassis of our bodies, the vehicles of our souls. So as the birthday party of the cosmos is just getting

underway ask yourself: when was the last time you have had a full service tune up? Had your divine spark plugs cleaned, been throttled by a powerful teaching that straightens your alignment and either inflates or deflates your tires as needed.

This Galaxy we drive is surely made in heaven but its maintenance is done here on earth. We are forever needing to tune ourselves, to scrape away the rust, the chipped paint tainted by driving ourselves to distraction causing occasional fender benders, not to mention those caused by drunk drivers on benders. But remember that we were given an unlimited warranty from *Hashem** and these regular trips to the body and soul shop are virtually free of charge at this very station.

So fill your radiator with the distilled water of life you've been given and let it radiate bountifully to all whom you encounter; make sure your distributor feeds the hungry, be sure to nurture your grow plug, and most of all make sure that what you have learned is given to your children in a great transmission, because it's not automatic. It's a manual transmission that requires lots of shifting on a bumpy road full of twists and turns.

* A creative way of not speaking God's name.

The Month of Elul

Also during the month of Elul Jews do self-scrutiny work, that includes looking into the mirror/s of the self.

Approaching the High holy Days gives this attention special resonance. At times we don't like what we see in the mirror both physically and spiritually. The critical self can get quite activated if we are truly being honest.

Yet, if we look carefully enough, we can find transformation and even redemption within those very findings. This, however, takes and extra measure of patience and perseverance, along with a dash of wisdom.

Would anyone call this experience fun? Well, let's stretch our imaginations a bit.

Slichot:
The Hall of Mirrors

Each year the month of Elul brings with it a circus inside my psyche.
Far from the big top is the dreaded Hall of Mirrors where I confront

my own grotesque and distorted face, my gaunt string-bean self
where I doubt that I am enough, the gigantic head reminding me

of how full of myself I can be, and of course the dark mirror,
the fear that I may be alone in the universe. Wondering why I've

dared to enter this house of horrors, I encounter a small room
filled with an overpowering white light and I am stunned. Now after

decades of therapy, meditation, Reb Zalman teachings, and lunches
at Saul's deli I realize this is no ordinary room but a personal Holy of

Holies and the force of that light has transformed those three dreaded
faces. The string bean self is now the fragile being that lives within.

The gigantic head, a call to do helpful acts for humanity, and the mirror
of darkness: a reminder that both light and dark comprise the rich

nutrients of cosmos. Every year this circus apparition returns and
affirms the truth that the face I see in the mirror each morning

is actually just an ordinary person, perfect in his imperfection, in this
fragile, spectacular and endless hall of mirrors.

The Magic of Letters

Ancient languages such as Hebrew and Sanskrit reveal meanings in direct and immediate ways. In English the word *denote* signifies to give an exact meaning, whereas *to connote* is to suggest or imply a range of possible meanings. Like hieroglyphs the Hebrew letters tell a story directly. Even their English cousins tell stories. The letter B (Beit in Hebrew) on its side demonstrates baby, born, begin, buns, and so forth. What we call *B'resheet* makes perfect sense for *in the beginning.*

The letters Beit, along with Resh, and Kaf, (BRK) in Hebrew call us to explore what it means to bless. The English-Hebrew dictionary says *to speak well of,* or *to worship:* while the Hebrew itself directly demonstrates the essence, literally *bending the knee at the well.*

If we have the discipline to trace the original meaning of the letters (regardless of whether or not we read Hebrew), new worlds of meaning open up. In Hebrew letters we encounter the crossroads where prose and poetry meet.

Brakhot:
At the Well of Blessing

To bless is to bend down

and retrieve water from the deep well of wisdom for
someone who's thirsty for much more than liquid.

To bless is to bend down into the deep well of caring
into the pain of a friend, releasing the ill humors
that can be expunged only by bending low enough
to notice downcast eyes and
offer the untainted droplets of empathy.

To bless is to bend on one knee at the ultimate well,
the one not made of stone,
and propose marriage to the holy one

whose answer will surely be yes and the nuptial celebration
will be euphoric but over the course of time that union
will become more challenging and require humility

because the marriage of this world and the other,
like all marriages is an arduous undertaking,

and can only thrive if we do these things
on bended knee. And by the way, let's not forget

about the deep well of justice. Many around the country
these days are *taking a knee* in front of that well too.

Remembering and Forgetting

Binary thinking would suggest that remembering is good and forgetting is bad. But we can think of instances in which remembering every moment we've endured is less than helpful and forgetting (forgiving) is a great blessng. Part of the work of repentance is to let go, forget to wear our masks, and to remember to do the work of forgetting.

Zackar — To Remember
Shakash — to Forget

The forty days of preparation require the human
heart to a challenging yoga: remembering to
remember and knowing when to forget.

The heart struggles and juggles this paradox,
stretching itself because remembering must
contain the act of forgetting.

I try to remember my pitfalls and failings.
I practice the art of forgiveness, that
brighter side of forgetting.

As the day of holy remembering approaches we
we will be passing through the door to encounter
what lies within,

because the primeval shore from whence we came
crawling is a holy calling, and with its yearly date
of expiration reminds us that the act of *T'Shuva** is

our heart's desire, to be held to the fire
standing before *Hashem,*** and for that pristine moment,
ever so sacred, we can forget our self-important tasks,

take off our foolish worn out masks,
and dare to be naked.

* Hebrew for returning.
** Hebrew for the great vibration/God.

Being Called

At times we feel called. This calling has been described in a variety of ways. These days one might say: "I've been called to volunteer at some organization that serves the poor." Or, "I feel a calling to start a new creative project." Maybe the calling is simply to slow down and enjoy each day.

The Torah, the Hebrew Bible, has a slant on this calling business that seems to be quite compelling however, something more than the activities described above. *Hineni* is Hebrew for *HERE I AM.* I put this in caps because this is a big deal. Or was a big deal? It seems that the likes of Abraham, Moses, the Buddha, and Jesus were called to change a people or a civilization.

These sorts of callings seem to have a few things in common. Speaking for the Old Testament I can say that (1) the person being called feels, at first, incapable of doing the task. And (2) they always seem to answer with this strange response *Hear I Am.* (3) The *Here I Am* comes first.

The order of these responses baffles me. If some person or cosmic voice were to call me out, and I heard it clearly, and asked me to run for governor for instance, I'd say, *are you kidding me?* but the biblical order of things is first *Here I Am* and then the considerations, the excuses, and the rest.

Why?

Hineni: Here I Am

is a koan. Your name is called, you don't say
what? or *who's there?* You say *Here I Am!*
Where did that voice come from?

Hashem calls to Abraham moments before his
knife is about to make the dreadful sacrifice.
And yes we'd be vexed if we lost the next one
in line to spawn our kind. But does it not
confound you how the sound of an angel's voice
can so transform a moment into paradise?

As for Moses, a voice calls his name and rather
than run from the sacred fire in the burning bush
his refrain is the same: *Here I am,* Knowing his
fears and speaking woes, those nagging voices say:
Who am I to fly so high? Yet Moses goes.

Hashem is always calling and those derisive
voices beneath our skin are the stupor we're in,
thinking we must be super to just be the human
star that we really are. And I suspect that *Here I Am*
is a divine reflex of a kind, a karmic infusion
that over-rides the trivial fears of the mind.

To recognize that voice is itself a daunting proposition,
so be ready for the moment when your name rings out
beneath the desert sun and you're the chosen one to
pay attention to an open door to a new dimension.

No, you can't be Moses or Abraham, but you can be
who you are and say . . . *Here I Am.*

Malkut

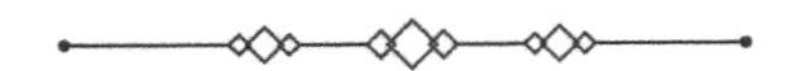

The word *Malkut* is heard many times during the High Holy Days. It is often translated as *kingdom* or *sovereignty.* Yet, in the Kabbalistic rendering of the tree of life, it lies at the bottom of the energetic chart depicting consciousness. We tend to think of king and sovereign as being above. A clue might be that *Malka* means queen. Perhaps there's a ruler or power that doesn't *trickle down,* as we might say these days. Perhaps sovereignty is a word that must be re-visioned, updated, downsized, and upended.

Everything seems to be up-side down these days in the *kingdom.*

Malkhut and the True Nature of Sovereignty

From some mysterious universe you signed up for this lifetime
committing to offer the world your unique and selfless gifts, not
out of the desire for fame or to please the relatives.

With the help of the infinite both brain and heart are wired and
ready for those moments wherein you are the one confronting
the bully, befriending the outcast or inventing the simple device to purify water.

You are the one learning to give with no strings attached but your god
given heartstring. This quality called *chesed, metta, karma yoga* or *agape*
seems trivial compared to the exalted acts of philosophers and kings.

All the while, however, the loving presence of the *Shekinah* has been
hovering nearby, artfully prodding you and me to practice the true
lowfallutin nature of sovereignty.

Remembering Body and Mind

Simply put, we have to be aware that we as humans are composed of both body and mind. Regardless of how inter-twined the two may be, the split can be quite formidable. A spiritual practice is necessary in order to overcome what has transpired over the course of millennia.

Ultimately we are neither solely our body nor our mind. But that's a different issue. Before we can really know who and what we are we have to play a bit with this split. Maybe in the relaxing and playfulness we will encounter some of what we are looking for.

Zichronot: Body and Mind

Your mind aches for memories of your mother's love.
The body remembers both the joy and the trauma.

You try to sit quietly to calm the mind.
The body has its own agenda.

Your mind keeps you up in the morning.
The body moans: "Go back to sleep."

Your mind craves to know who birthed the universe.
The body wants to dance with it.

Your mind believes that it is the crowned jewel of creation.
The aching body busts that illusion.

You finally hear the voice
that truly remembers and it says:
"This contentious marriage
has been going on for decades.

It's time for you two to get it on".

Starting the Year With a Bang

As a child the holiday Rosh Hashanah was the Jewish version of the American culture's New Year's celebration.

Like the secular holiday we had a version of New Year's Eve, and of course New Year's Day, which was the end of vacation and a mixed blessing of sorts.

But most significantly, it was, well, *New Year's,* a simple celebration of the next year, and as such, had no real cosmic flavor. We didn't talk about creation or the planets. Some did say *the world's birthday,* but this was rather in fun, a creative way to welcome the new year.

I'm not clear when things changed for the Jewish people. Maybe for some (the mystics) it was always about the beginning of things, but we had no real clear idea when things started—except how it was expressed in Genesis, which seemed obscure and in any case contained several different versions of *the beginning.*

Along came the Kabbalists a few hundred years ago and then science and now *cosmology,* and along with those changes, a whole new way of relating to the significance of Rosh Hashanah, literally, the head of the year. It's much more than that.

The First Birthday Party

Mother watches as her son Cosmo is about to blow
out his first birthday candles. He takes a deep breath,
tummy contracting, eyes widening, cheeks billowing,
then an ominous rumbling sound begins bellowing

as brilliant lights start making a cosmic quaking,
the biggest-bang-iest never-been-always-will-be explosive
shaking beyond time commences and shattering shards
of light go scattering everywhere rocking the empty shell

of nonspace as the first cosmic wind blows a ten billion-year
firestorm through the party hats, cupcakes and balloons
and oh what a confusing and delicious mess then ensues
that melts in your mouth and in your uplifted hands,

and now eons later we're still drinking the Kool Aid of our
petty distractions while planets whirl and birth moons,
oceans fill up and empty out, insects birth larvae, reptiles
birth hatchlings, mammals, their young, and we still stumble

and bumble with our theological mumbo-jumbo, all the while
forgetting that we live in a cosmic gumbo of blinding light,
pregnant dark, sunset rays, mysterious days, and boundless praise.

The neighbor asks, "How will you pick up the broken pieces from
from this cosmic shebang?" We pause for a moment and then say,

"Through acts of kindness, justice, peace, and forgiveness."
The neighbor then asks, "Won't that take a long time?"
We say, "Yes, a very long time."

Seeking and Letting Go

One of the more powerful rituals of Rosh Hashanah is called *Tashlich.*

During the afternoon, after formal services, groups of worshippers find their way to a body of water—a stream or pond or even the ocean—to embody one's errors or misdirections from the previous year, or beyond, and allow them to be carried to the metaphorical *ocean* where all such supplications can be received and transformed.

For many, this ritual act supplants all of the others performed during the Jewish new year. For me *Tashlich* is even more meaningful when I consider those with whom I am performing this ritual. It matters to me, in ways that I can barely articulate.

Still, I will try.

Malkut and Seeking

I want to be with those who seek what is secret,

I want to amble beside sacred waters with those who
toss their sticks and feathered oak leaves downward,

along languid streams carrying flaws that reveal precious
learnings, and hearts that reveal lifetime yearnings.

I want to dive headlong into the watery underworld
with comrades at my side who are drunk with the divine,

who dance and play, drum and pray, and are
drenched with the tears of grieving friends
right there in nature's forest pews.

.

We'll become divining rods that discover the secret wells
that navigate their way downward into the great river

that sings the great song and hey, the drinking
at that watering hole goes on all night long.

Shofrot Will Rock Your World

Blowing the shofar is not just about announcing the new year; it's a mindfulness meditation, a wake-up call that I need in order to remember. Unlike the alarm on my phone or the old-fashioned alarm clock, this startling ear-blasting ceremony will shake you to the core. Long before "We will, we will, rock you," there were *Shofrot*.

Shofrot: The Moment of Awakening

Wake up! The shofar is sounding, you are connected
at the hip with the deep azure sky
that you thought was a painted backdrop.

Wake up! And eat that breakfast muffin of kindness
and the sweet melon of earthly delights
that welcome the morning.

Wake up! And cherish the friend or partner nearby,
for you are not the lone stranger
on this wild spinning planet.

Even the barista at the café might be
the unwitting emissary of a morning blessing
for *Peete's* sake, or a spark that will ignite a

new adventure of the soul so wake up
and smell the coffee and the mint
and the breathtaking roses you so often ignore.

For God's sake, wake up and flip an otherwise droll
September morning into a
brilliant necklace strung together by

beads of epiphanies, and remember that the shofar's blast
is spirit's trusty alarm clock
joyously ringing and continuously

singing to get your attention even when you hit the
mundane snooze button of your life.

Today, maybe even right now: Wake Up!

Shofars: Announcing the Creation of the World

I really like the idea of the blowing of a ram's horn announcing the creation of the world.

It could be a siren-like a hurricane warning but that would signal some terrible event. It could be a news article or a text that reaches everyone's phone, but the banality of that idea sickens me. It could be word of mouth, but that would take forever and next year's celebration would have already passed.

The sound of the ram's horn is wild and shocking but also playful, funny, and alarming to an extent—and the horns are animal parts and smelly and twisted into elaborate shapes, colors and sizes and they have various musical pitches. Yet one need not be a musician to master the art of playing them, and children can do it and they love it and gather eagerly to be a part of the ceremony. And of course adults have always done it and continue to do it and then dance and drum and get pretty riled up in a way that seems related to creation itself. You really ought to try it once, at least.

Shofars: Things Great and Small

Back then the ram's horn welcomed the new moon, when
when creation herself was expectant in the dream time.

Back then Hashem appeared at Mount Sinai as a thick cloud
by day and fire by night, encapsulating sacred time.

Back then the shofar's piercing sound caused the Hebrews to
tremble in awe, and the silence that followed stopped time.

Today as then, the shofar welcomes the expansive sun and
and the contemplative moon, reminding us that we still live

face to face with our shamanic past, human lips pressed
against twisted animal horns, urging us to revel in each

blessed moment where breath meets sound, spirit meets
flesh, and life winks at death. It drowns out the futile

sights and sounds that feed our distracted minds and points
us toward the naked, rip-roaring and feral truth that despite

our folly and despair, some force greater than all of us here,
is always there.

Is It Closing?

Religious rituals often take a life of their own; that is, what starts as a powerful and innovative connection to the rhythms of life and change (the words rhythm and ritual come from the same root) over time can become rote.

The gate is closing, the book of life will be shutting, in short, *you'd better hurry and do your penance or you will be left behind.* Some think you might die in the next year based on how you negotiate the last hours of Yom Kippur, the day of atonement.

The Gate is Closing

The ship will soon embark; those waiting to pass through the gate
with its laden doors are eager to get on board. All eyes watch as it
moves, all ears listen as it creaks, and hearts are
beating fervently as it is about to close.
But not all hearts.

There are mothers in the kitchen too busy making the evening meals.
There's a man who has cared for the hungry animals, and a woman
who is attending to a sick friend.

Those rescuing children at the border will miss the boat. Those
teaching kindergarten in the Fruitvale are simply too
exhausted to race through the gate.

Those offering comfort to the dispossessed have forgotten with their
minds but their hearts are attending to a world crying out to be comforted.

Show me where in the Book of Life I should concede that
they will be excluded. Show me rather, a greater source, toward a gate
whose lock has long ago rusted away, and the
glistening signpost at the entrance says
we are always open.

*Baruch Hashem,** you and I will be inscribed into the book of life.
And to those whose prayers and supplications
are lost, delayed, or unspoken may you not feel bereft.

Even if the ship carrying the book of life has sailed,
and even if you believe you've been left behind on the island of death,
know that you've been spotted,
and in God's time your ship will come in.

* God willing.

Sukkot

In the present century many of us are fortunate enough to have what we call *permanent dwellings*. Actually of course, there is no such thing, in practice, it's sometimes wise to meditate upon the impermanence of all people and things.

In this regard I have always felt a special fondness for the Jewish harvest festival known as *Sukkot*. Temporary huts are built, near or connected to the house, but always with loose construction and partially open roofs of willow leaves and palm fronds so as to view the night sky, the planets and the stars.

Sukkot comes towards the end of the High Holy Days each year. It is the fifth day after Yom Kippur and feels like welcome relief. Instead of gazing inwardly at one's flaws we spend time gathering the fall harvest and gaze upon the heavens that seem to embody perfection itself.

During Rosh Hashanah and Yom Kippur we clear out the parts of our inner lives that no longer serve us; during *Sukkot* on the other hand we welcome the bounty of the physical and spiritual harvest that is always ours for the taking.

Sukkot: Gimme That Old Time Religion

The herbs, fruits
and vegetables of the season
are not grown within the city gates,
because we are farmers, spirit farmers,
each year enacting a diaspora, gathering a
harvest that is not only food for the body.
We rise and fall like the bread from the earth,
like the grains out in the field we dance and wave,
unaware that our jaunty display will suddenly, one day,
be cut short. Atop our structures the fissured spaces
within palms and willows welcome the stars of the
night sky, reminding us that the cosmos still
sparkles and that shaking the lulav to the
six directions is Hebraic shamanism.
On these days we honor both worlds
and linger in what is so often
ignored and forgotten:
less house, more sky,
less sun, more moon,
less work, more being,
less doing, more sitting,
In this paradox of natural bounty and emptiness
we begin to see that we human beings are all transient
and undocumented caravan-walkers passing
north or south,
to and fro, willy-nilly, upon this blue-green
fructiferous earth.

Chapter 7

The Hebrew Bible: Brief Stops Along The Way

Going Forth

There's a big deal both in Genesis and in basic grammar about the word go. Avram (whose name will become Abraham) is commanded to *go forth,* a common translation of the phrase *Lech Lecha.*

Scholars have been investigating, tweaking, revising, and reconsidering this phrase for a long time. If I were your mentor or therapist and I thought that you were avoiding your calling, or afraid to live and love and risk, I would probably be saying something like: hey, get going!

I'd quote the mystic Rashi who said: *Go for yourself.* Forget what the world needs for a moment; do it for yourself. Doin' your own thing may actually be exactly what the planet needs, but that's a whole other enchilada. Hearing the calling seems to be the more salient point. It was a ground-breaking point in the movie *Field of Dreams* when Kevin Costner's character heard a distinct voice continually whispering: "Build it and they will come."

We have so many ways to fall into our comfort zones—food and drink, entertainment, and social media. We seem to have our attention almost entirely on what's up in the airwaves, what's up in politics. *What's up with you?* we tend to ask.

In ancient times, it seems that gods or goddesses or angels delivered direct messages to people without the internet, therapists, self-help books, or even the priests or rabbis of which I speak. One Hasidic story (I can't recall where I heard it) tells of a disciple asking the *Rebbe* why G-d no longer seems to speak to us humans directly as in biblical times. The *Rebbe's* response: "Because these days we don't bend down low enough."

Lech Lecha: Abraham and Folks Like Us

Some folks are called to make miraculous journeys.
Abraham and Sarah were told to leave home
and go to a strange land,
change names and bless others. They complied.

Before blowing the whistles and horns think about your and my ancestors from Spain, Prussia, Poland or Morocco heeding the same call.
They had huge garage sales, packed up the rest of their possessions, took a boat, changed their names
and became sources of blessings
for their generations to come.

We too might get a little love for discarding outworn theologies, discovering ourselves by moving west, changing our names, (temporarily to Ocean or Durga Das), inventing new occupations, bending gender, smashing idols, reinventing sacred dancing, and drumming right there on the bima!

If we've learned anything at all, it is that if we can
see the path that we are on, then we are not on it.

Like Abraham and Sarah, we were called to get
out of *Dodge,* go forth to a land we know not, and
accept an assignment we could not fully comprehend.

Abraham and Sarah are not prophets to be worshipped,
but examples to emulate.

Sitting Between Worlds

We humans sit between worlds. Safety is perceived to be inside our tent. Mystery is often in the faces of total strangers. We might pause from our direct confrontation with a revered authority and bake bread for those strangers who could be delivering tidings of joy. Whoops! Wrong religion. A great *B'rukha* (blessing) that we could hardly imagine.

Come to think of it, three strangers or wise men or angels appear in various traditions. Maybe the number three has some significance that we should consider.

Actually, all numbers, stories, and the details in those stories have much to teach us. As we stroll into some of the *parasha* (weekly Bible readings) and holidays not as famous as Yom Kippur, let's keep our eyes and ears open. With a bit of aging gray brain matter, we can take in these stories in a whole new way that can be as warm and nurturing as the bread that Sarah made for the strangers (angels) who travelled to the now famous tent to tell Abraham and Sarah that they were to birth a child, a nation, and several new religions.

Parasha Vayeira

At times you sit at the edge of your tent, balanced precariously on
a precipice of dreams, seeing what lies within and beyond.

An apparition appears, a bright light, and you believe that you
are conversing with, well, the source.

Yet when three strangers appear you to run towards them and
prostrate yourself, welcome them, give them water, feed them,
and attend to them under the cover of a great shade tree.

They ask, "so where's the wife? The husband?" You say: "behold, in the
tent." You realize they have come from another world to break the
news that you are to birth new life, a people, a legacy for the
millennia.

You thought you knew love and compassion, but in this instance some
voice inside said: "excuse me while I attend to strangers." Now you've
taken the Jewish Bodhisatva vow.

Both you and I are the hands of divinity.
Outside of our familiar tents
there lies a bright empty space filled with angels
whose faces radiate a
wink and a nod as they test the breadth of our hearts.

In the blinding heat of noon, sit quietly and listen for
that great presence
outside your door. But don't ignore the hungry strangers,
who might be
bringing the next miracle that lies in store.

About Isaac and Stories

Abe and Sarah just learned (from those three strangers/angels) that Sarah would give birth to a child the following year—when she was 90 years old and Abe was 100. The next thing we know, their child Isaac is a teen I think, and the infamous and imminent sacrifice of the lad is stopped miraculously by another one of those angels. They sure are busy and helpful.

Once again, let's be watchful that, as we delve into these myths as I just called them, we realize that just as in dreams, there are numerous facets to observe in these diamonds. Notice what grabs you and realize that the powerful details were meant to grab different people at different ages, with different experiences, and certainly at different stages of psycho-spiritual life.

Binding Isaac

A Torah study group is meeting and dutifully wrestling
with the parasha: *The Binding of Isaac.*

In the back sits a man who recently lost his beloved son.
He sees no meaning
in this horrifying death sentence of Isaac.

A new mother in the room cries out, shocked by the cruelty
of a divine entity testing the faith of Abraham this way.

A scholar of history sees the end of the era of sacrificing
firstborn sons to a callous god, and a woman approaching
death sees guardian angels.

Those in the room who are grieving
see the face of loss, the victims of cruelty,
more cruelty, and the dispirited are disheartened.
These *midrashim* are all real and true.

Beyond seeing, however, is something greater that can
only be known in the realm of dreams and visions.

The binding of Isaac is a hot swirling desert wind from
the east bearing the next inscrutable mystery. Squint-eyed
and struggling, we yearn to understand,

but what we purport to know is narrow, like *Mitzraim.*
There is seeing, but as Abraham said, *there is vision.*

Jacob the Moon Child

In the Old Testament there are a number of major heroes. Obviously Adam and Eve, Abraham and Sarah, Job, Moses and Miriam, even Pharoah. At the center of it all is Jacob. He's one of the early patriarchs, but much more. Adam by contrast, is a player in one of at least two creation stories, but we know no so little about him. He's not a person that we can relate to. He's an archetype or symbolic of the ground, the earth, manifestation, or who knows what else?

Jacob, however, embodies what I see as the heart of Judaism. Esau, his older brother (twin), is a blip on the screen. He's here and then gone because he's what some would call the solar twin: the aggressive, hungry, impetuous hunter, an old school guy who'll trade his birthright for a bowl of porridge. Jacob, however, is the *lunar* guy who is connected to his mom more than his dad. After Jacob does the birthright trade and tricks the aging Isaac into giving him the eldest son's blessing, Rebekah, his mom, warns him to get away and the story line of Judaism is then off and running.

You probably know the rest; Jacob is the famous dreamer who wrestles the angel and births the name Israel. So in a real sense he's the prime mover of the story of the Jewish people and, of course, the two major religious traditions that follow.

Take a careful look at the storyline of the Hebrew Bible and you'll see that Jacob's progeny become the protagonists of The Book, from his son Joseph (Pharoah's dream interpreter a.k.a.—shrink), next to Moses and then Joshua and prophets, and on and on. Jacob is the sensitive twin, the dreamer, who leaves the hunting tent and strikes out into the wilderness where Judaism and indeed western culture itself start to find an identity that is unique in the world narrative.

Jacob, then, deserves a great deal of our attention, and it behooves us to delve into the psyche of a man embodying a tale of twin brothers, two parents, two wives and two lives.

Jacob's Days and Nights

Daylight can be quite misleading. Before Jacob knew Rachel
or Leah darkness was his consort in that room where he, the
gentle lunar twin gestated in the heal of Rebecca's womb.

Then, years later with Isaac's stolen blessing, twilight again
beckoned Jacob to the dreamtime where hobnobbing angels,
visiting him in the harsh land, scurried up and down, and up
again to the pinnacle of the ladder's celestial crown.

We see life's daytime dramas unfurled on his face.
Emblazoned with the battle scars of servitude, competing
wives and sons who will lead infamous lives.

In the barren land, once again, darkness accompanies
his fleeing, angelic wrestling with that nameless being, thigh
sinew, and bone—a numinous naming ceremony to return home.

The daylight of our lives can be quite misleading.
The secret name is only revealed between the murky waters of
the Galilee and the Dead Sea, where light is muted and shadows

speak the lone decree, between the land of holy thou and illusions
of a separate me, in the netherworld where we cannot see: but
there we sit in the still and fertile darkness, called to dream and

struggle, we are blessed with our true name so we can fully be.

About Joseph and Dreams

We know of cultures in which dreams are the centerpiece of village life; one such people, the *Senoi* people of Malaysia, are reported to have no crime or mental illness. What we know for sure is that Jewish culture has always honored the wisdom of dreams, from Joseph to Freud and beyond.

Often people fear their dreams because of nightmares or so called pre-cognitive dreams that seem to foretell dangerous or tragic events.

Having studied and taught dream work for twenty-five years, it is my experience that by far, the most common dream is a symbolic nod from a relatively unfiltered core element of self that wants to illuminate significant parts of one's inner life, including glimpses into the soul's path.

There may be no Bible story more popular than the story of Joseph: his dreams, his brothers, his coat of many colors, and his journey to Egypt that precedes that of his entire tribe. This story is an integral prelude to the Passover story.

Let's have a bit of fun with it.

Joseph at the Well

Maybe you have a Joseph inside, within your hide,
and your dream of grandeur is dressed in a brilliant
coat of many colors that awakens others' fears
of their own dreams denied.

And so you are sold or so we're told
by those heartless brothers, but don't deny
that you too have your own jealous brother,
your critical mother and wounded other,
your own captain of Pharaoh who is your
your CEO of narrow.

They too reside, deep inside.

And though your own youthful dreamer
has heartbreaks in store,
they are surely of the mysterious kind.
In times of dread, your own winged dreams
of bowing and kow-towing sun, moon, and stars
illuminate some sacred thread that
binds you to the One Great Mind.

When you're cast in Joseph's pit
and carried off to a land below, there is
still that presence that you used to know.

So that one day you will stop and think,
at your magic well you'll pause and drink,
and ponder how such dangerous dreams
and twisted schemes led you
to be Pharaoh's shrink.

About Joseph and Goblets

Some say that there's nothing new under the sun and I'd add, there's nothing new under the silver moon also.

The light we seek is ever-present and we cling to the illusion that we invent it every century or two, often in the form of some abstruse liturgical text. We delude ourselves and imagine that we have invented the wheel, birthed the divine child, and discovered the holy grail.

Look more carefully and notice that in the story of Joseph and his brothers is, yes, a silver goblet. We claim the latest version as our own proprietary invention, forgetting that an empty cup, the youngest child, the color silver, and an impassioned calling are the elements that make the narrative divine.

We seem unable to remember that a goblet holds space, that the color silver reflects the perfect warmth of the sun, that the youngest child often gets the toughest assignment, and that some visionary being might be plotting this whole befuddling business. We are experts at forgetting.

Joseph and the Silver Goblet

Jacob went up the ladder. Joseph goes down into Egypt.
Your spiritual travels take you to heavens and underworlds.

Joseph was enslaved by his brothers and went to prison.
You've had your family betrayals and personal traumas.

Joseph scales the ladder of success in Pharoah's domain.
You've had your fifteen minutes of fame, maybe more.

Joseph's brothers came to Egypt after fleeing the famine in
Canaan. Eventually he planted the silver goblet in young

Benjamin's bag of rations. Even in his high station in Egypt,
he saw himself in his youngest brother for Joseph too, as a

young and brash child, dreamed that the moon and stars were
bowing to him. The goblet is an invitation for Benjamin, and

for you and I, to dare to live in the world of outrageous dreams
and visions. The silver vessel is lunar alchemy, the Jewish

Grail, and has been placed into your bag of rations by a hand
greater than you can imagine. The *parasha** called *Mi-Ketz*

concludes with the words: "Only he or she in whose possession
the goblet was found shall be my slave; the rest of you go back in

peace to your father." To those of you called to Joseph's world,
welcome fellow slaves.

* Weekly Torah reading.

Tu'Bishvat

The subject of light seems to be quite central to understanding our place in the universe. We could approach the subject by studying the physicists over the last 500 years: Christiaan Huygens in the 1600's, James Clerk Maxwell in the 1800's, and of course Max Planck and Albert Einstein in the 20th century. Put simply, light behaves as a wave if we have the tools and patience to scrutinize it.

Light also of course has magical qualities: a *now you see it, now you don't* quality. There are frequencies of light that we can see and others that we cannot. In that sense it's a bit like the light we talk about in psycho-spiritual language.

Follow the light, love and light, or even "I am the light." Clearly, there are dimensions to light that humanity has explored for millennia and most spiritual traditions have much to say on the subject.

In Judaism we have a mid-winter holiday know as *Tu B' Shevat* (literally the 15th day of the month of Shevat), the very day I am writing this essay. As children, we were told that this odd holiday was the birthday of trees and, as such, we should plant them—and the state of Israel,then, was the preferred location.

Politics and religion were conflated in curious ways, both then and now. As I've explored the mystical side of my Jewish tradition, I've come to see the universality of this holiday that honors the magic that lies underground in midwinter. It gives birth not only to the trees of the emerging spring, but to all beings. We have silent, dark places within that give birth to life—if we have the wisdom and discipline to pause. Even in California, winter calls us to look inside.

Tu'Bishvat in the Suburbs

Some days I wish that the whole electrical power grid
would just shut down.

We'd watch the sultry late afternoon creep gently into the blush
of a rose-colored sunset where we'd greet our neighbors standing
just feet way on our suburban lawns, and be awestruck as the sky's
inky canopy begins to enclose us, where we'd break into spontaneous
rounds of applause as both we and the sun are swallowed up.

With no radio, TV, internet, or lights of any kind, we'd see the crimson
crowns of the red-headed woodpeckers with their checkerboard wings
stopping to drill and tap at the mighty Modesto Elms on our block,

competing with the clicks of crickets and the howls of dogs
and the squeals of children in utter amazement, with the elm's having
the last say as they whisper their reminder that humans, birds,
crickets and trees are all children of the same natural order.

We'd feel the light behind our eyes and
bathe in the light glowing inside our yearning hearts—
and we'd celebrate the midwinter gloom
knowing that the sparks of light gestating
beneath the pregnant earth illuminate
much more than a groundhog's refuge,

and we'd eagerly embrace our place in the world
as the radiating filaments of divinity that we are,
woven into the fabric of this great star-embroidered
cosmos and we'd continue to do all of these things

until the electricity comes back on.

Next Year in Jerusalem

This is a phrase that most Jews living outside of Israel know and espouse during the days of Passover. Those living in the Holy Land need not utter those words, literally because they reside in proximity to Jerusalem.

But, of course, taking religion literally is a trap for sure, so what do we make of this salutation or blessing?

The Jewish diaspora, from the Greek word for scattering, has become a common term of reference because, since the destruction of the second temple in Jerusalem in year 70 of the common era, we've gone, among other places, to Babylonia and the Middle East, Turkey, Spain, North Africa, Poland and Russia, the Americas and beyond. *Next year in Jerusalem*, for many or even most Jews, means *may I be in the holy city next year* during the Passover celebration.

I know for me, and for most of those around me, the phrase is a metaphor for what is happening within heart and soul at this special time of the year, the time of spring and renewal.

Passover is a nearly pure celebration of social justice, and the wish that all peoples will be freed from bondage of all sorts. Some, I'm sure, see it as merely a tribal holiday and, as such, more exclusively about Jews. I certainly hope that the tribalism that now plagues America and elsewhere will not keep us from seeing the global picture that astronauts see from space.

If we are at all awake, we can study history a bit or look at those photos of the beautiful blue-and-white planet swirling through the galaxy with no borders or boundaries visible. A look at that might just be called *Olam-shalem,* a . . . world of peace.

Of Hollow Places and Hallowed Ground

Each Passover we Jews say: Next Year in Jerusalem!
Not this year.

The city of peace resides in the hollow center of my chest
that is too often filled with anger and judgment.
Better I should stuff mindfulness notes
into the cracks of my own wall
than google Travelocity for tickets.

And, yes, I will chant the *Dayenu's* knowing there is no end to gratitude,
but the great thank you is less about Egypt and deserts
than this hallowed moment that will be lost
unless its fragrance is savored like *charoset* and sweet wine.

The beauty inside all living and breathing beings,
the plants, animals and the shimmering stars in heaven,
is not merely a day in the life of Moses marveling at a bush.

It's the daily hallowed practice of firewalking with new eyes.

And the freedom you and I prize is hollow until
all who dwell upon this hallowed ground called Earth,

enjoy its bounty.

"I've Seen the Promised Land. I May Not Get There."

Martin Luther King in his last speech sensed that he personally wouldn't make it to the promised land. The calling is from a voice that is most often not heard. It comes from a place that is completely mysterious and must be so much less about you and I in particular than our ego selves can understand.

Dr. King, for reasons we both know and for reasons at which we can only guess, chose Moses's journey through the desert as the guiding light of his great vision for the civil rights movement. It might not be a bad idea to revisit Moses for a bit.

Both he and Dr. King have things to teach us.

Moses Strikes the Rock

You and I and Moses, all drawn from the Nile's waters
No one exempt, excluded, turned away or thrown into
cages at the Texas border.

Moses does his duty, fights oppression in Egypt, kills a man,
flees to the desert, meets his wife, receives a great calling,
gets drafted to become the MVP of the great Passover,
and leads his tribe through challenging waters toward a
promised land but cannot enter. Because he strikes a rock?

No, not crime and punishment or sin and redemption,
it's not a god or angel missing the three-pointer at the
buzzer. Striking the rock twice is a call to pay attention
and take the reins ourselves. It's strength in numbers.

We the people complete the shamanic journey upwards
and back to the land of Abraham and Sarah.

What gushes forth is the water of life carrying us on our
journey of accepting our calling, crossing turbulent seas
and offering portentous teachings.

No salvation by proxy, we're all on duty.

Moses, like us, is of two worlds, resistant and bumbling,
shamanic and divine.

He doesn't enter the promised land but he does fulfill
his promise. Likewise, we will not receive a ticker tape
parade into Canaan or know where this journey will end.

Arms greater than ours bestow life, lift us up and guide us
toward curious destinies. But you and I must strike the rock,
often more than once, before a place in the desert will be
crafted to receive the body, ready to take its rest.

Business, Shabbat, and Choices

Shabbat, the Sabbath, the day of rest, call it what you will. It's a day when we suspend normal work activities to the extent that we can and give over that time to matters of spirit and contemplation. What all of that entails is of course a complicated matter. Some people adhere to strict rules, *religiously,* while some honor the sabbath rituals occasionally, and others just enjoy a day off for leisure.

What is significant in my opinion is what lies at the core of this gift that grew out of Judaism and now is celebrated by much of the world. Whether following ritual practice or being by oneself, the jewel of this practice lies in what is invisible. We can chant the chants or earnestly recite the prayers but unless we find that core of releasing the constraints of the work-a-day world, we have missed a great opportunity to just be.

This is a difficult yoga to practice much less master. It takes some muster, some intention, indeed it's a compelling choice.

Business as Usual

There's business as usual, the drama
of family trauma, things that lurk at work,
bills missed and that tyrannical to do list.
There's also a time and place that simply
is.

At such moments, I notice the bare branches
of our Japanese Maple tree undressed by
December's rain, showing off its spindly
arms of glossy crimson.

That tree, called a *Sango Kaku*, reaches up
to its sacred mountain for blessings of light,
and has roots that meander downward to the
place where the truth of things is nourished
by the plain dark earth. That tree simply
is.

You and I however get hinged to deadlines,
commutes, and daily fiascos. Business as usual
becomes a viscous liquid that fills the vessels
of our lives, a run-on sentence that will
never auto-correct.

Fortunately our traditions give us a day of rest.
More than a time without working, Shabbat is
kin to the Sango Kaku tree, quiet, reverent,
unscripted, and blessed with tranquility for a
dazzling twenty-five hours, all in the welcoming
arms of her cushiest lazy boy chair.

That time simply is, a gilded invitation to
re-enter the web of perfection that is, that
was, and that will always be despite
business as usual.

So act now because this special offer is limited.
Tonight as *Shabbat* approaches
will it be business or *is-ness?*

Chapter 8

Love and Its Mysterious Manifestations

Our Confusion About Love

It's my opinion after decades of thought and readings, that love is the most misunderstood concept, with the possible exception of god. The service element of love that we witness in Mother Theresa pushes us in a helpful direction, as does the social justice work of a Martin Luther King or a Gandhi.

I am certainly one who gets caught in the trap of those familiar and *hifalutin* ways of dealing with acts of love. Over the decades the less-flamboyant ways of being with others certainly speak to me, and yet, truly being mindful of them in my daily walks and interactions is still challenging and elusive.

We all want a good and comfortable life. We all want to have *those favorite things*. Some of us however seem to excel at noticing that others don't have any things and those who do notice have a knack for acting in wonderfully selfless ways.

I've been noticing more and more that this form of loving carries with it some discomfort. There must be a magical yoga going on that some have mastered. Maybe I discount some of the things I do naturally that don't seem so special. Yes, love is confusing.

This Confusing Thing

An otherwise drab fall evening is unfolding and
you are chopping carrots and onions for the soup
that will sustain the two of you for the week.

The next afternoon at the café you patiently listen
to the friend whose predictable complaints cannot
be comforted, but the feeling becomes the healing.

At the morning service you reach out to the woman
you barely know to inquire about her grief, and later,
you hold the gaze for the lonely man on the walking
path whose heart seeks a mere hello from your eyes.

Are these moments all bound together in the same
sheaf as diamonds, roses, and orchestrated proposals?

Is the soup pot, the generous ear, and the empathic look
less worthy of the love moniker less than the glistening
exploits of your romantic life? Where does love reside?

We toss around words like messiah and conjure up a man
in sandals from a distant past or a hopeful future when
actually, the messianic age lies in the very next generous
and unscripted act that you stage.

Heart Transplants

During the High Holy Days we also focus on the connections that our hearts enjoy with other human beings. For the most part we assume that *heart connections* are referring to the energetic and spiritual realms.

Science however has recently proven to us that we are connected in ways that you will find quite amazing.

Sharing Hearts, Beer and Junk Food

A woman in New England received a heart transplant, awakened from the operation, and immediately felt hungry, craving foods that she had never before desired: beer, Snickers bars, and Chicken McNuggets, the favorites of the donor. (no life style comments please)

Another woman received a heart transplant and awakened feeling furious only to learn that her donor had died while fighting in a sports bar. These factual events rightly give us pause.

They cause me to crave nuggets of Albert Einstein's heart mind, Nelson Mandela's heart courage, Maya Angelou's heart wisdom, and a host of others who radiate patience, forbearance, compassion and loving kindness.

I'll avoid the medical route god willing, and try the cognitive-behavioral, as well as the transpersonal-spiritual approach. Still, you and I can share enlivened heart molecules with our fellow human beings through caring eyes, open ears, attentive dreams, and maybe we'll even try flying over the rainbow.

So take it, take another little piece of my heart baby and I'll take a piece of yours, and we'll forgo the Chicken McNuggets and I'll learn what makes your heart tick and what makes it skip a beat and you'll respond in kind by feeling the ache in my heart and learn what is my heart's desire and we'll stop this nonsensical mind chatter and have heart to heart talks with each other, our children, our friends, even our foes.

One day we will remember and give thanks to the one who originally started planting hearts in creatures and human beings in the beginning and realize that the yearning of the universal heart is a world-wide and joyous conspiracy, the one and only original and ongoing purpose of the whole cosmos . . .

So chew on that rather than the McNuggets.

Hearts and Symbols

Sometimes I remember a quotation from the works of Alexander Pope: "A little knowledge is a dangerous thing." This sentiment is often attributed to Shakespeare, where it actually reads, "A little learning," but neither version addresses what I consider to be the salient point here.

Over the last few centuries so much information, theology, philosophy and learning have penetrated what I call the mass mind, that as a species we may actually know less about what is essential than we did while hovering in ancestral villages. At least back then we were speaking a language we all understood in a context that made sense.

Some theologians suggest that ancients utilized symbolic thinking more than we do today, given recent centuries where what is literal seems to have orchestrated a hostile takeover of myth and religion.

The subject of love was not and is not immune to this happenstance. The symbols that we choose and the images that represent those symbols have in mass culture become ubiquitous. I imagine that in today's world, be it in New York, Tokyo, France, or South Korea, Valentine's Day is celebrated with hearts and chocolate and represents romantic love.

It takes some scratching and digging to find meanings and traditions that penetrate further.

Last Valentine's Day I found myself trapped within this very dilemma.

Valentine's Day Symbols

It's Valentine's Day and I'm pondering. What offering
might be the right expression of love for my wife?
I'm befuddled.

On the radio a Stanford scholar informs me that long
before the famous sculpted red heart we all know,
ancient peoples revered the pine cone.
I'm chuckling.

I do more research and find images of Osiris, Shiva
and even Popes holding staffs topped by pine cones.
I'm intrigued.

I learn that the pineal gland, known as the seat of
the soul, derives from the word pine, whose flower
pre-dates all other flowering plants.
I'm excited.

I realize that the snakes of the Greek Caduceus, the
Mayan Chicomecotl and Hindu snake deities circle
and bow to pine cones atop icons of higher knowledge,
spiritual insight and love far beyond hearts and chocolate.
I'm transfixed,

but still befuddled about that Valentine's gift.
So I give up, buy a dozen crimson red roses, feeling
ashamed of this prosaic act until I realize that the rose,
like the pine flower and the pineal gland, demonstrates
the Fibonacci sequence of all beings on planet earth.
I'm speechless.

Then preparing my midrash poem for the next sacred
gathering, I am reminded that "Jacob gazed upon the
face of god" and hence, "called that place Peniel."
Now I feel a calmness in the center of my being.

And by the way, Audrey just loved the roses.

Love's Underbelly

I believe you are starting to see that these love poems are largely about what I call *love's underbelly*. Furthermore to go on to invoke the phrase *spiritual context* is to suggest that in the greater picture, love may be an endeavor far beyond the usual way books, movies and theologies often describe it.

Forgive what may sound trite or overstated, but so many teachings point toward the possibility that love is the central issue for humans on planet earth. One wouldn't report that by observing the behavior that extraterrestrial beings would see, reporting back to their homelands, so there must be a formidable gap between most earthly activities and this true purpose.

These essays and poems attempt to investigate the little-understood and underappreciated ways that what seems painful or hopeless on the surface, may reveal an underbelly of meaning and transformation that supersedes what we blithely call love.

If this is true, then we are obliged, I think, to explore both the vagaries that normally describe love and the little-appreciated nuances that exist, even if they seem unpleasant, or downright ridiculous.

And I do not claim to have some special or esoteric knowledge that is crying out to enlighten humankind. I am hoping, however, to pose some compelling questions and to offer a few humble insights.

Love Hides

Love hides in crevices that go unnoticed.

It dangles from the worn threads of faith
that drop from religion's coattails.

It lies in splinters, beaten by the club of family strife
and the slow decay of relationships.

It stares wistfully from outside shattered windows
of illness and mortality.

It sits amidst the debris left behind in silken cobwebs
by spirit's door and in breathless moments
when the body can do no more.

It even lurks in moments of anger and hating.

Love hides in crevices: unperturbed and waiting.

Oy! Loving One's Neighbor

Is there any biblical proclamation more famous than this?

Both Bible testaments advocate this pretty clearly but I wonder if anyone really loves another as one loves oneself?

Also then, who is your neighbor? The guy next door of course—but how about the desperate guy who breaks into your home?

And of course loving your enemies gets mentioned: If you're really practicing loving everybody, then why would you even have enemies?

Who really has the capacity to love another person who embodies what you despise? Do you believe it when so many espouse this great aphorism, or is it a snow job?

I realize that I might start to drive myself crazy with all of these questions that clearly have no answers so I'll come at it another way.

Love Thy Neighbor As Thyself

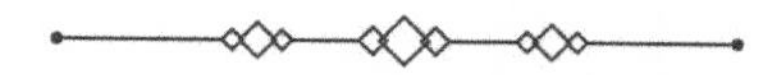

Really? Are you kiddin' me?
I don't even like my neighbor's dog.
A trick question for sure.
I struggle to love family and friends at times.
Love thy neighbor as thyself!

Come to think of it, there are so many
parts of me that are embarrassing,
broken and worse that loving my neighbor
as myself could be counterproductive.

Who wrote the little biblical gem?
It sounds so simple, like Nancy Reagan's
"just say no."

I realize that I'm stuck in a prickly conundrum.
Clearly I'm supposed to look into the mirror
and ponder all of this.
To love him or her or them, I've got to love me.

Maybe Fritz Perls or Carl Rogers concocted
such Jewish Koans. Maybe Koan and *Kohanim**
are related.
Perhaps this assignment is supposed to throw a
psycho-spiritual wrench
into the center of my being.

If so, it is succeeding. Loving one's neighbor
as thyself is not a sudden realization on a
mountain top but a humble admonition to get
started by trekking up that mountain
with an empty rucksack,
a broken heart, and the courage
to open to the next revelation.

* The priestly class in the Jewish tradition.

Loving A Tree?

Hey, that tree over there and millions like it abound. I know that I ignore them all too often. When I do bother to notice, I'm immediately awestruck by, forgive me, their sheer size. This sounds embarrassing I realize but if you stop and take notice, at least on our walking path, there are these tree creatures that are larger than elephants or blue whales.

Some are larger than houses, and yes they are composed of wood and delicate leaves, just imagine. Their trunks wind around like great serpents and their high branches seem to know exactly how to reach upward and yes what I am saying is obvious but why we don't feel more awe every time we gaze upon them is beyond me.

On top of all of that, there are hundreds of them nearby and thousands more scattered around and millions of them literally everywhere and if nothing else decent happens in the next weeks or months, or even years, well, at least there are trees.

Loving a Tree

Please,
stop for a moment because
I would like to ask you something.
Have you ever considered the possibility
that it might be life-altering to really notice a tree,
and even better to plant one like our Japanese Maple
just feet from our front door. In mid-winter its five-fingered
leaves have withered, revealing smooth shimmering red branches;
we gasp and wonder if those tinted glowing arms have always been
so or did neighborhood children paint them red in December or
have we simply ignored the generous gift of their wintry
window display reaching upward toward
our nearby mountain where today
the enlivened heart of some Moses
or Myriam might be awestruck
witnessing the work of great
hands that once struck
a stone with a cosmic
match and ignited all
living things with a
fiery presence that
is forever visible
if only we could
remember to gaze
less with our eyes
and more with our
wide opening hearts.

Chapter 9

The Unholy Alliance of Politics and Religion

It Happened

I did not want to write the poem on the next page. Particularly on my birthday. Particularly on October 27, 2018.

Pittsburgh, Pennsylvania.

Evil usually unfolds in fits and starts more than explosive bursts.

When it feels close to us or is close to us, it is an explosive burst.

I am heartbroken and for this tragedy and so many like it . . .

It tends to happen on a day like today.

On a Day Like Today . . .

We are feeling the
first glint of shock our
ancestors felt the day they
were expelled from Spain.

Maybe now our restive hands
are sensing the first drops of
pelting rain that fell on loved
ones who boarded an
unspeakable train.

We remember those who wore
yellow stars—and perhaps those
times are not so far away.

Maybe soon on a day like today
we will see crescent moons on
the sleeves of those who have
no place to pray.

Glance upwards at the angry and
ominous pall in the sky casting
unforgiving clouds over the frightened
heads of brown-skinned children
who are pleading: Why?

The latest version of *it can't
happen here* is no longer
news from a distant
shore; it's here at our door.

My hope is that you and I will
awaken and be vigilant on
behalf of all that we hold to
be dear,

And my prayer is that our tired
eyes can see and our broken
hearts can hear.

The Bible and the Flag

One of the more perplexing phenomena in this era is the alignment of the so-called Christian right in America. *Jesus forgives* is the response of choice to people and events that normally would not be tolerated.

Will we look back upon those affirming the politics of 2016-2019 as we do some Germans who allowed the Nazis to come to power? Comparisons of this nature have seemed out of the bounds of normal political discourse. But to many it looks like a duck and winks at cruelty and brutality like a duck, and excuses actual Nazis with torches in the streets like a duck. It used to be hidden just below the surface of the political conversation. No longer.

Today is five days after the slaughter of eleven Jewish worshippers at the Tree of Life Synagogue in Pittsburgh, Pennsylvania.

"When fascism comes to America it will be wrapped in a flag and carrying a cross." (Sinclair Lewis: American novelist and playwright, and winner of the Nobel Prize for Literature)

Two Images Come to Mind

January 30th, 2016
A candidate in Iowa holds up a Bible like a trophy.

June 19th, 2018
A fake president goes to a rally in Tampa and hugs the flag.

"When fascism comes to America it will
be wrapped in a flag and carrying a cross."

Who knew it would happen so
literally?

Keeping it Real

Sometimes you hear the expression *go for the light.* Also perhaps, *love and light.* Light seems to be the hero. When we do therapy, take workshops, or even attend concerts that enliven ourselves, it's easy to revel in our own growth processes. Some spiritual teachers have coined a term for this sort of inflation: *spiritual materialism.*

Yes, it's wonderful to be blessed enough to focus on the higher mind, the so-called *winged life*, and the ongoing search for meaning. At the same time, we hear: *'Tis the gift to be simple.* The world goes on even after an uplifting concert or a meditation retreat. The mind's ability to judge some activities as worthwhile and others as beneath us seems limitless.

At times we need a little dose of hard-nosed blue-collar reality to keep things in their proper perspective.

McDonald's

Never had this word sullied a page of my text until last
last night when our world music ensemble presented
Congolese rhythms, Ladino loves songs, Brazilian
Sambas, and Machado's poetic wizardry put to music.

After this onslaught of musical lovemaking I was driving
a friend and cohort home. Said he: Let's do a small errand
on the way and stop at McDonald's, "for pie." Having not
dined at one since 1959, I was taken aback. Could this be

a joke? No, but this was no ordinary evening. My friend
lives near the venue in question in Oakland California
where a midnight snack at McDonald's might not be so rare.
Realize that I'm white, straight, and these days, suburban.

He's gay, black and urban, but we both celebrate the music
that emerges from its seed source. We do however share
the job titles of healers, seekers, facilitators, and partakers
of escapades contrary to the America of neon lights, fast

food, and fast minds fed by tech monsters and starved by
mass culture. Tonight, however, after sampling the tastes
of world music bundled with dancing bodies, it was indeed
the moment to cross another momentous chasm of this day

with a tiny apple pie in a cardboard sheath along Berkeley's
storied Telegraph Avenue. Now, I strive to cross barriers
of mind and heart and celebrate spiritual diversity. Yet if
I am to truly cross interfaith worlds of spirit I must strive

to transcend the mindset that elevates me too near Icarus
and his daylight flight toward the sun that dwells all too far
from the all-too-challenging world of the tattooed man
who handed us our minipies at the drive-through.

My hope is that while he keeps his eyes on the pies, he can
still keep his eyes on the prize.

About a New Exodus?

It's sometimes difficult to avoid the trap of telling these stories without believing them to be historical facts, and thereby missing their universal significance. Let's hold the story of the great Exodus out of Egypt without seeing Charlton Heston's face or an imaginary newspaper clipping with maps and charts of the crossing. Leave that to the History Channel and the televangelists hawking religion as verifiable truth rather than the wondrous and mysterious entity that it can be.

Any people escaping from tyranny and danger can be *The Exodus.* We don't have to search too carefully to find examples, yet some who cry fowl at others' sacred passages have, no doubt, forgotten about their own. I'm often mystified when fellow Jews rail against immigrants fleeing death in Central America who show up for sanctuary at our southern border. They seem to ignore their own people's recent history, including those who fled Hitler, only to be turned back by this country and sent on boats back to Germany where they perished. How that similarity can be so readily ignored is a painful mystery to me.

Exodus: The Never-Ending Story

Exodus
from America
is a phrase that one
is beginning to hear more
and more during these days.

The E word
sits in our DNA.
Yes, we'll celebrate
the great escape from Egypt
but it's never what is seems in the
land of those *Mitzraims**: Canaan, Rome, and
Spain, tragic illusions and heart-breaking dreams.

Yes, we'll
celebrate spring and
renewal, the miracle of
creation, but along the tracks
of our pilgrimage we have had our
original tickets punched merely as travel
visas, affirming the truth that all beings on this
earth are undocumented immigrants walking hand

in hand
through the
the Sinai sand,
across the Edmond Pettis
Bridge and the parched desert
darkness toward the Rio Grand.
The hands that penned the Torah did not
begin with the creation or the fall but clearly
proclaimed the duty for us all to heed the next call
when the next Pharaoh starts to build the next wall.

*The Hebrew word for ancient Egypt/the "narrow place" both in geography and the psyche.

Good and Evil

Too much has already been said about good and evil. I won't offer platitudes or the usual profundities that appear in history, novels, or political discourses.

On the other hand, it's too late. We are surrounded by challenges to our way of life and our democratic institutions. Even the most basic sense of justice and decency seem to have been eroded to a dangerous degree.

I can only imagine what some of my heroes would have thought if they could see what has happened to America in the 21st century. They certainly stared evil in the face; they might have faced greater challenges than what we are facing. I can only acknowledge that what I feel at this moment in history seems quite dire.

One of the most compelling questions human beings face concerns the issue of good and evil. I often wonder if in the aggregate one outweighs the other.

At the Flea Market

Last week at the flea market
I spied Mahatma Gandhi,
Rabbi Abraham Heschel, and the
Reverend Martin Luther King
perusing a small two-pan measuring scale.

One pan was marked good, the other pan, evil.
A discussion then ensued.

Said Heschel: "this scale is flawed,
the opposite of good is not evil,
it's indifference."

Gandhi replied: "yes, I agree, for good
and evil often are found together."

Then Dr. King spoke:
"I find this scale to be be befuddling
because 'there's some good in the worst of us
and some evil in the best of us.'"

With that they simply walked off.

I timidly stepped forward and bought the scale.
I took it home and measured the weights sitting in
the two pans marked good and evil and here is what I found:

When compared, good and evil seem to be
about equal in measure,
but clearly, at times like this,

it's necessary to put a finger on the scale.

Chapter 10

Death and His Cronies

Corfu and Legacies

I'm guessing that you, like myself, might wonder if what we do in this short lifespan has any impact beyond our days. I suspect that we have little control over such matters, and rightfully so. Our assignment, I believe, is to live our lives fully, live the lives we were meant to live, live good lives – and these are but of few of the aphorisms that address the subject.

Still, what we create in this life does speak volumes about the life that we are living, and the word *creating* itself puts us in alignment with the very cosmos. Most humans don't have the knowledge or wherewithal to invent a life-saving medical device or overthrow an evil dictatorship, but we all can, I believe, take on the small adventure that lies before us, be it planting a tree, rescuing a kidnapped goddess, or writing a book.

Whether this constitutes a legacy or not is out of our hands. Better we should attend to what is at hand.

Olives, Myths and Words

Barely shadowing my parcel of sunlight overlooking
the Ionian Sea with her placid azure waters are two
diminutive olive trees, silvery green counterpoints

bent like an aged couple facing off, gnarled and twisted,
roots exposed, pockmarked and struggling to stand.
Who plants trees knowing they will bear no fruit for
a dozen years?

Eons pass and Menelaus's kidnapped wife launches
a thousand ships, kings and warriors battle for a
decade, Paris, Achilles and thousands more die.

Another decade unfolds, this drama an underworld
of sirens and sea monsters as the Odyssey bears its
narrative fruit for generations.

What Olympian gods orchestrate such a drama wherein
myth and history embrace as do the olive, and the tree
that births it?

In our time the British authors Lawrence and Gerald
Durrell descend into the crystalline waters of Kalami Bay for
future readers and scholars hungry to partake of word
and verse.

They had no titles and few prospects.
What beings plant such seedlings for fruits knowing
that they would only be gathered posthumously?

Knowing how fruitless would be the self-indulgent grasping.
Knowing that creating and even nurturing reaps no instant
reward. Knowing that with olives, myths and words,

there is all the time in the world.

Peppers and Aging

There are numerous metaphors about humans who are *aging*, itself a euphemism for getting old. We used to just say it. He's an old man or she's an old lady.

Now we soften the blow with words like aging, getting older, getting along in years, even *maturing*.

If I'm blessed with reasonably good health as I enter my 70s, I certainly hope that I will still contain the spark that motivates me to live a life that is exciting, or possibly even uproarious.

There seem to be more and more examples of such folks these days. I hope to be among them.

Padron Peppers

They say that eating Padron peppers is like playing
Russian roulette; most are mild but some are wickedly hot.

They are quick fried and blistered while still maintaining a
shimmering green exterior, so that, lying in the pan they
appear to be tiny uniformed soldiers crowded together.

You never are quite sure which one might jump up and
bite you, lead an insurrection, start a coup d'état or even
a full-scale rebellion.

I hope that I won't lose the Padron pepper inside of me.
When I find myself lying in the neat unformed rows of
the docile and aged, I pray that I will have the strength
to speak my mind, agitate and instigate, drawing on my

spicy-hot riotous and rowdy rebel who will burn the
tongue right off of a world afraid of its own shadow.
That will turn up the lights a notch!

About the (Four?) Seasons

We talk about death, fear it, write about it, investigate it, and it has always been central to the issues of living in its specter. Now something new has arrived.

It's no longer just about our personal mortality, or the death of loved ones, one's tribe, religion, or country. Of course it's about the planet, whose grim future so often gets put on the back burner of our busy lives.

Are we trapped in the dilemma of climate change having been politicized to the point of no return? Are we headed toward an unlivable planet?

The answers to these questions are frightening and unforgivable. It is too much to hold the enormity of all of this. Human beings have never faced a challenge this catastrophic, and yet we forget, deny, postpone, rationalize, pray, recycle, and grasp at whatever small strategies we can.

I feel embarrassed to even be writing these words as if you might be uninformed, which I know you are not. As the next shiny political object is waved in our faces, what we face is still incomprehensible. Someday soon we will bury our communal face in our hands and wonder what else we might have done.

Vivaldi and Monsanto

There are no more woods in our woodlands.
We've committed treason against trees.

No more evening cries of crickets in
thickets or reasons for the four seasons.

Fall's loamy carpet has fallen away,
Brutal summers linger longer each day,

Winter's snow falls only
in window stalls at malls.

Spring's annual floral display is now
poisoned by Monsanto's death-march decay.

We've bequeathed our planetary future wishes
to swarms of cockroaches and jellyfishes.

The words I offer you today have
rhythm and rhyme but there's no
more time

and nothing hopeful
to say.

After You're Dead

Most days we simply live life. That may include stumbling to the bathroom, stretching, remembering and writing down a dream, preparing for work of some kind, doing errands and relating to other human beings in a variety of ways.

More and more, as I'm aging, I find myself wondering what my own death and passing might feel like to my loved ones. My wife and I are blessed to have been assisted in creating a living will, but that only takes care of the business of dying.

What I'm exploring here are the messages that I want to communicate to my children that are more about precious items, both material and emotional.

One place where these two worlds collide is indeed: stuff. Most people I know have far too much of it. One evening some friends of ours joined us in considering what it might be like for our children to go through our things, when that day comes, and imagine what they'd feel at such moments.

Reasons Your Children Will Love You After You're Dead

As they rummage through what they anticipate will
be reams of yellowing graduate school notebooks and
discover that you discarded them:

They will love you.

When their own child is vomiting as they are leaving for a
long-awaited concert, they will recall similar moments in
their own childhood and

They will love you

When they realize that you've persevered
through personal differences
such as omnivore vs. vegan, introvert vs. extrovert,
and OCD vs.
post-cataclysmic modern:

They will love you.

And, when they are doing their own inner work and realize
that you attended to your own, thereby freeing them to
soar higher and dig deeper,

They will love you . . . even more.

About Synchronicity and Death

First of all, everything is connected. We sense that and know that. But when death strikes and we find ourselves below or outside of the linear world where we normally reside, a strange sort of magic occurs. Carl Jung called it *synchronicity*. Yes, it relates to timing. As a drummer I often say: "timing is everything."

I especially experience synchronicity while traveling, being outside of the normal rhythms of life. I run into distant relatives and friends I haven't seen in decades. Things just seem to fall into place, or some force seems to ordain certain events.

In my experience, these sorts of moments happen quite often when death strikes. Why? I'm not sure. Maybe we are more sensitive to the connected nature of things; maybe a magical atmosphere surrounds us and causes the unusual and the bizarre to unfold.

Quiche and Kindness

It was bound to have happened at last night's poetry salon. Lovers of the oral tradition trickled into this cozy pastel home in central Oakland. What most caught my eye was the bird feeder, just outside the window, that kept our winged friends busy scurrying as we humans dabbled and pecked our way around our winged life.

Then *he* arrived, a graduate school compatriot I hadn't seen in 33 years, since that other evening when he had brought two homemade quiches to our door. He had entered that evening, with visible fear and uncertainty on his face, given these were no ordinary circumstances. You see, my wife and I were sitting shiva, given the all-too-recent death of our stillborn son Jacob.

Last night's encounter revived that distant memory, as Proust's famous almond cookie had. He took the only open seat (next to me, of course) where memorized soul poetry was about to be recited randomly, spontaneously, and as suddenly as death itself.

There was and is nothing anyone can do in such circumstances but show up, quiche in hand, eyes soft and uncertain, body language contrite, demeanor wrapped in sack cloth, performing a simple act of kindness for a couple mired in the bottomless swamp of infant death.

He didn't remember cooking or delivering the manna when I recounted this story, nor did that matter. We remembered every detail of his act, ordained by a force not of this world, replayed by those who received, at a moment when no healing was yet possible. Greater minds than ours crafted this moment, greater hands than ours carried the ceremonial offerings, and a greater heart than our broken hearts crafted this ritual that still evokes salted father's day tears that are now 33 years running, and counting.

Where Is Heaven?

This heaven business. In Hebrew it's called *Shemayim*. As I understand those root words: *shem* is sound/vibration and *mayim* is water. The Books of Moses really starts with water and darkness, then wells of Abraham, and then the Red Sea and so forth. I believe that this is all metaphoric of course, about something more than desert wells.

If heaven signifies anything, it must be about the vibratory realities of that which makes us most human and most caring about one another. In that sense, then, heaven must be closer to us than we often realize. And the question of *where* must be as meaningless as *when*, since quantum science has taught us that time itself is malleable.

Add these elements together and one might conclude that heaven is less a distant reality and possibly something more omnipresent.

Prayer for Rain

There are times, difficult times, when we gaze upwards
towards God's heavenly sky. But what if we concluded
that heaven was really more nearby?

What if heaven were beneath our feet, where we
walk along troubled streets filled with sidewalk signs,
exposing desperate people in desperate times.

What if Heaven resides in our eyes? In the well where
tears arise, and we deeply see the blank despair of those
who merely seek, in vain, their share of this earthly prize.

What if heaven's rains will only come cascading down
when our hearts do truly yearn for what is under that
crackling canopy of thunder.

Perhaps it's breaking hearts that spawn the rain and
touch the hand of the one who opens *red* seas, parts
the watery depths, and with joyous torrents of medicine-

rain, invites all who dwell upon this earth to cross the
pregnant swelling Jordan River, not just one nation,
but all who are welcomed into the vast embrace

of her creation.

About the Author

Bruce Silverman M.A. is a poet, musician, ritual maker and a transpersonal counselor in private practice. He founded and directs the *Sons of Orpheus* Men's Community and Healing Arts Center and has taught world drumming and dream exploration at Holy Names University, Matthew Fox's Naropa University and throughout the United States. He and his wife, Audrey Silverman Foote, founded *Shir Neshama,* a Jewish Renewal Havurah in the San Francisco Bay Area. His website is www.brucesilverman.org.

CPSIA information can be obtained
at www.ICGtesting.com
Printed in the USA
FSHW011133120319